AF352826

Transitioning to
Good Health and Well-Being

Transitioning to Sustainability Series: Volume 3

Series Editor: Manfred Max Bergman

Volumes in the series:

Volume 1: Transitioning to No Poverty
ISBN 978-3-03897-860-2 (Hbk);
ISBN 978-3-03897-861-9 (PDF)

Volume 2: Transitioning to Zero Hunger
ISBN 978-3-03897-862-6 (Hbk);
ISBN 978-3-03897-863-3 (PDF)

Volume 3: Transitioning to Good Health
and Well-Being
ISBN 978-3-03897-864-0 (Hbk);
ISBN 978-3-03897-865-7 (PDF)

Volume 4: Transitioning to Quality
Education
ISBN 978-3-03897-892-3 (Hbk);
ISBN 978-3-03897-893-0 (PDF)

Volume 5: Transitioning to Gender Equality
ISBN 978-3-03897-866-4 (Hbk);
ISBN 978-3-03897-867-1 (PDF)

Volume 6: Transitioning to Clean Water
and Sanitation
ISBN 978-3-03897-774-2 (Hbk);
ISBN 978-3-03897-775-9 (PDF)

Volume 7: Transitioning to Affordable and
Clean Energy
ISBN 978-3-03897-776-6 (Hbk);
ISBN 978-3-03897-777-3 (PDF)

Volume 8: Transitioning to Decent Work
and Economic Growth
ISBN 978-3-03897-778-0 (Hbk);
ISBN 978-3-03897-779-7 (PDF)

Volume 9: Transitioning to Sustainable
Industry, Innovation and Infrastructure
ISBN 978-3-03897-868-8 (Hbk);
ISBN 978-3-03897-869-5 (PDF)

Volume 10: Transitioning to Reduced
Inequalities
ISBN 978-3-03921-160-9 (Hbk);
ISBN 978-3-03921-161-6 (PDF)

Volume 11: Transitioning to Sustainable
Cities and Communities
ISBN 978-3-03897-870-1 (Hbk);
ISBN 978-3-03897-871-8 (PDF)

Volume 12: Transitioning to Responsible
Consumption and Production
ISBN 978-3-03897-872-5 (Hbk);
ISBN 978-3-03897-873-2 (PDF)

Volume 13: Transitioning to Climate Action
ISBN 978-3-03897-874-9 (Hbk);
ISBN 978-3-03897-875-6 (PDF)

Volume 14: Transitioning to Sustainable
Life below Water
ISBN 978-3-03897-876-3 (Hbk);
ISBN 978-3-03897-877-0 (PDF)

Volume 15: Transitioning to Sustainable
Life on Land
ISBN 978-3-03897-878-7 (Hbk);
ISBN 978-3-03897-879-4 (PDF)

Volume 16: Transitioning to Peace, Justice
and Strong Institutions
ISBN 978-3-03897-880-0 (Hbk);
ISBN 978-3-03897-881-7 (PDF)

Volume 17: Transitioning to Strong
Partnerships for the Sustainable
Development Goals
ISBN 978-3-03897-882-4 (Hbk);
ISBN 978-3-03897-883-1 (PDF)

Antoine Flahault (Ed.)

Transitioning to
Good Health and Well-Being

Transitioning to Sustainability Series

MDPI • Basel • Beijing • Wuhan • Barcelona • Belgrade • Manchester • Tianjin • Tokyo • Cluj

EDITORS
Antoine Flahault
Institute of Global Health,
University of Geneva,
Switzerland

EDITORIAL OFFICE
MDPI
St. Alban-Anlage 66
4052 Basel, Switzerland

For citation purposes, cite each article independently as indicated below:

Author 1, and Author 2. Year. Chapter Title. In *Transitioning to Good Health and Well-Being*. Edited by Antoine Flahault. Transitioning to Sustainability Series 3. Basel: MDPI, Page Range.

ISBN 978-3-03897-864-0 (Hbk)
ISBN 978-3-03897-865-7 (PDF)
ISSN: 2624-9324 (Print)
ISSN: 2624-9332 (Online)
doi:10.3390/books978-3-03897-865-7

Contents

About the Editor

Antoine Flahault MD, PhD in biomathematics. He was appointed as full professor of public health in 2002. He was the founding director of the French School of Public Health (EHESP, Rennes, 2007-2012), co-director of Centre Virchow-Villermé for Public Health Paris-Berlin (Université Descartes, Sorbonne Paris Cité), co-director of the European Academic Global Health Alliance (EAGHA),and president of the Agency for Public Health Education Accreditation (APHEA).

He conducted research in mathematical modelling of communicable diseases, chaired the WHO collaborative centre for electronic disease surveillance, and coordinated research on Chikungunya in Indian Ocean (Inserm Prize, 2006; was scientific curator of a large exhibition Epidemik, la Cité des Sciences et de l'Industrie (Paris, Rio and Sao Paulo)). In January 2014, he was appointed as professor of public health at School of Medicine, University of Geneva where he is the founding director of the Institute of Global Health. He was elected corresponding member at Académie Nationale de Médecine (Paris). In January 2014, he had more than 235 scientific publication referenced in Medline.

Contributors

BENJAMIN J. KNOX
Associate Professor, Researcher Center for
Information and Cyber Security, Norwegian
University of Science and Technology, Norway.

CLAS-HÅKAN NYGÅRD
Professor, Unit of Health Sciences Faculty of
Social Sciences, Tampere University, Finland.

COLIN BINNS
Emeritus Professor, School of Public Health,
Curtin University, Australia.

EDUARDO MISSONI
Professor, Department of Policy Analysis and
Public Management, Bocconi University, Italy.

FLORIAN FISCHER
Dr., Institute of Public Health, Charité –
Universitätsmedizin Berlin, Germany.
Institute of Gerontological Health Services and
Nursing Research, Ravensburg-Weingarten
University of Applied Sciences, Germany.

FRANK J. CHALOUPKA
Professor, Institute for Health Research and
Policy, University of Illinois at Chicago, USA.

FRANZISKA CAROW
Dr., School of Public Health, Bielefeld
University, Germany.

GIUSEPPE CARRUBA
Dr., Division of Research and
Internationalization (SIRS) ARNAS-Civico,
Italy.

JANE SCOTT
Professor, School of Public Health, Curtin
University, Australia.

KAIE MAENNEL
Dr., Researcher Centre for Digital Forensics
and Cyber Security, Tallinn University of
Technology, Estonia.

MATTHEW CANHAM
Research Assistant Professor, Institute for
Simulation & Training, University of Central
Florida, USA.

MI KYUNG LEE
Senior Lecturer, College of Science, Health,
Engineering and Education, Murdoch
University, Australia.

RICARDO G. LUGO
Associate Professor, Institute of Psychology
Inland, Norway University of Applied Sciences,
Norway.

STEFAN SÜTTERLIN
Professor, Faculty of Health and Welfare,
Østfold University College, Norway.

SUBAS NEUPANE
Adjunct Professor, Unit of Health Sciences
Faculty of Social Sciences, Tampere University,
Finland.

VIOLETA VULOVIC
Dr., Institute for Health Research and Policy,
University of Illinois at Chicago, USA.

Abstracts

Global Health Determinants and Limits to the Sustainability of Sustainable Development Goal 3
by Eduardo Missoni

The Agenda 2030, signed by the Heads of State and Government in 2015, set out 17 Sustainable Development Goals (SDGs) and, for each of them, a number of targets to be reached within the next 15 years, with a total of 169 targets. SDG 3, "ensuring a healthy life and promoting well-being for all at all ages", provides, among others, Goal 3.8 "achieving universal health coverage, including protection of financial risks, access to quality essential health care services and access to safe, effective, quality and affordable essential medicines and vaccines for all". Agenda 2030, unlike the global agenda for the previous fifteen years, which focused on the so-called Millennium Development Goals (MDGs), mainly concerning the poorest countries, involves and commits all governments to the adoption of "indivisible" and universal goals that will help to end poverty by 2030 "once and for all", and also brings the issue of development back to a global dimension. The~new agenda is not without contradictions. Among other things, it proposes, among its economic objectives, "sustainable, inclusive and sustained growth", an oxymoron that was pointed out at the beginning of the 1970s, when the Club of Rome showed the "limits of growth" in a finite ecosystem. Thus, the~challenge of sustainability is global and involves all national health systems. Using Universal Health Coverage, SDG 3's target, which seems to attract most of the attention, as the main focus, the~paper argues that SDG 3's feasibility and sustainability is highly dependent on transnational determinants which, if left unregulated by appropriate global governance processes, may jeopardize its attainment. Global determinants (international macroeconomic policies, migration, climate change, market forces, technological innovation, etc.) affecting health system functions (stewardship, resources generation, financing and the provision of services) are identified, and their impact analyzed. The~analysis provides suggestions for the identification of an urgent paradigmatic shift to ensure the effective sustainability of SDG 3 in general, and of universal health coverage (UHC) in particular.

Transitioning to Good Health and Well-Being: The Essential Role of Breastfeeding

by Colin Binns, Mi Kyung Lee and Jane Scott

The world is facing unprecedented public health challenges, including changes, due to climate change, that will affect every continent, and limitations on global food and water supplies together with an increasing epidemic of non-communicable disease. The United Nations Sustainable Development Goal 3, "good health and well-being", includes targets for child mortality, maternal mortality and reducing chronic disease. Breastfeeding is one of the most cost-effective public health interventions available for countries at all levels of development. In the first year of life, appropriate infant nutrition (exclusive breastfeeding to around six months) reduces infant mortality and hospital admissions in infancy by 50% or more. Breastfeeding followed by the introduction of appropriate complementary foods at around six months establishes a healthy microbiome, has a long-term association with reduced rates of childhood illnesses, hospitalizations, obesity and later chronic disease, while improving cognitive development. It is consistent with the historical cultural practices of all societies. While the development of infant formula has been of benefit to some infants, its inappropriate promotion has resulted in a decline of breastfeeding in recent decades, and, as a result, recent health gains in many countries have not been as great as they could have been. Formula use increases health care costs through increased illnesses and hospital admissions. The specific target for non-communicable disease in SDG3 is to reduce premature mortality from chronic disease by one third by 2030. The Global Burden of Disease project has confirmed that the majority of risk for these targets can be attributed to nutrition related targets, and it is estimated that annually 15 million people between the ages of 30 and 69 years die from a non-communicable disease (NCD), the majority from cardiovascular causes. Healthy life course nutrition, beginning in the first 1000 days of life, is a major public health priority in answering this challenge. Breastfeeding, exclusively for about six months, and appropriate complementary feeding, establishes a healthy developmental trajectory. Children are the population segment most susceptible to the effects of climate change, bearing an estimated 80% of the increased burden of disease associated with climate change. The health benefits of breastfeeding will provide some protection against the health effects of climate change, including projected increases in some infectious diseases. Increased breastfeeding will avoid the high environmental costs associated with the production of infant formula, including the use of large quantities of potable water and energy. In contrast, breastfeeding, as well as being the best infant feeding intervention, has a very low environmental impact. An important part of the sustainable development agenda should be to promote breastfeeding and its benefits and to reverse the inappropriate promotion of infant formula.

Taxation of Tobacco, Alcohol and Sugar-Sweetened Beverages for Achieving Sustainable Development Goals
by Violeta Vulovic and Frank J. Chaloupka

Premature death and disability rates from non-communicable diseases (NCDs) have been rapidly increasing with potentially enormous implications on global economy. Three major risks factors for NCDs are tobacco use, harmful alcohol consumption, and poor diet. National governments can create incentives for behavioral changes in the consumption of these products through fiscal policies, such as price increase through taxation. A substantial body of empirical evidence suggests that a significant increase in prices of tobacco and alcohol products through higher taxes can discourage consumption and reduce their negative impact on the individual's health and the economy. Emerging increasing evidence on the impact of consumption of sugar-sweetened beverages (SSBs) suggests the same. Consumption of tobacco, alcohol and SSBs can undermine development and attempts at meeting the Sustainable Development Goals (SDGs), including health and well-being, hunger and poverty alleviation, quality education, economic growth, and reduced inequality. The empirical evidence shows a strong causal effect of consumption of tobacco, alcohol, and SSBs on NCDs. The attributable estimated economic costs can be very high, and significantly higher than the tax collection on consumption of these goods. Moreover, consumption of tobacco, alcohol, and SSBs can crowd-out spending on food and nutrition and increase hunger. In addition, the attributed NCDs can lead to lost productivity and earnings, as well as high out-of-pocket medical expenses for their treatment, putting an individual and their families into a vicious circle of poverty. At the same time, children may be forced to drop out of school to take care of ill family members or to work to make up for lost earnings, denying them educational opportunities and having an intergenerational impact on their and their country's economic development and growth. This paper reviews the empirical evidence on the impact of consumption of tobacco, alcohol, and SSBs on SDGs to make a strong case to support a significant tax and price increase to help achieving a country's development goals.

Transforming Regional Agrifood Productions to Challenge NCDs—From the DiMeSa Study to the PASSI Project and Beyond
by Giuseppe Carruba

Today, the crisis of the agrifood sector across several geographical European regions, including our own, combined with social, economic and health inequalities at regional, national and community levels, is characterized by extremely critical aspects, mainly residing in the limited innovation potential of companies and small/medium enterprises (SMEs), the lack of integration with public–private research institutions, the insufficient systematization and organization of the existing resources in an extended territorial networking. This results in increasing difficulties of SMEs to exist in both domestic and foreign markets with characteristics of quality and competitiveness. On the other hand, several epidemiological studies clearly indicate that Western countries, including Italy, are witnessing a dramatic phenomenon consisting of an increasingly large prevalence of chronic noncommunicable diseases (NCDs), including cardio-, cerebrovascular and respiratory diseases, diabetes, obesity and cancer, whose causes are primarily related to (removable) lifestyle risk factors, notably diet. Based on this two-sided consideration, promoting both the sustainable consumption and production of traditional food products in regional, domestic and international markets—through a series of activities aimed at increasing their health and/or nutraceutical potential, to clinically validate their effects on both health and chronic disease(s), and to enable the rapid technological transfer and development of either processes or products—would represent a systemic strategy of high impact in the short, medium and long term for important expected outcomes from an economic, technological and healthcare standpoint. Lessons learned from the Dieta Mediterranea e Salute (DiMeSa) study and its recent advancement, the PASSI project, with both published and unpublished results, are presented and discussed here.

Impact of Public Health and Sustainability of Global Health Action for Achieving SDG 3
by Florian Fischer and Franziska Carow

Global health is an area for research and practice that places a priority in improving health and achieving health equity for all people worldwide. Therefore, it corresponds with the aims of the Sustainable Development Goals (SDGs) for providing universal health coverage and leaving no one behind (SDG 3). Public health is understood as global health on a regional or local level. Whereas "old" public health focused mainly on the prevention of infectious diseases and the healthcare for specific subgroups at risk, "new" public health takes a holistic and interdisciplinary perspective on aspects related to health and well-being. Inter alia, public health genuinely deals with the reduction of health inequalities and risk factors---not only by prevention on a behavioural level, but also on a structural level, such as environment and society. Thus, we understand health as part of each of the 17 SDGs. In a world characterised by globalisation as well as epidemiologic and demographic transitions, the impact and sustainability of global health action needs to be taken into account to improve the world population's well-being. Therefore, this contribution highlights present and future challenges for global health. Furthermore, it provides overarching suggestions and policy advices for improvements to strengthen a global response on today's challenges and to gain the targets set within the SDGs, by pointing out that almost all of the 17 SDGs are relevant for health-related issues.

On the Relationship between Health Sectors' Digitalization and Sustainable Health Goals: A Cyber Security Perspective
by Stefan Sütterlin, Benjamin J. Knox, Kaie Maennel, Matthew Canham and Ricardo G. Lugo

The healthcare system's efficiency and effectiveness depend on policies and procedures, such as the rapid and goal-directed exchange of health-related information between its different actors. In recent years, more national healthcare systems approached sustainable health benefits for their populations by digitalization. Despite numerous advantages such as the improved availability of medical services, their improved quality and cost-effectiveness, new vulnerabilities threatened the trust in the healthcare system. Healthcare institutions are a primary target for cyber attacks and have the stigma of being a particularly easy prey for malicious actors, due to the relatively low levels of problem awareness and preparedness. We argue that cyber threats pose a new challenge to reach sustainable health goals and have to be considered an imminent part of all healthcare development strategies. Cyber threats in the healthcare sector can have detrimental acute effects in times of national crisis, hybrid warfare, or international conflicts above or below the threshold of war. In peacetime and in western democracies, however, breaches of data that were processed by private or public bodies undermine the public trust in these institutions. Public awareness and scepticism therefore influence policies around the digitalization of healthcare, and this consequently affects the development of institutions at the frontline of healthcare and achievement of sustainable development goals. The chapter further discusses the relationship of working towards sustainable health goals, public trust and data security with the example of the Estonian e-health system, where leading principles such as cyber security, transparency, and patient's data autonomy are prioritized.

SustainableWork Ability during Midlife and Old Age Functional Health and Mortality

by Subas Neupane and Clas-Håkan Nygård

Background: Sustainable work ability is a multifaceted concept that involves the matching of the needs and abilities of the individual with the quality of work. Good work ability during a work career is one of the potential indicators of sustainable work ability and employment, as it requires a good balance between individual resources and work demands. We aimed to study the developmental pathways of work ability during midlife until retirement and its impact on functional health in terms of mobility limitations in old age using longitudinal data on employees in a large amount of blue- and white-collar occupations. Furthermore, we studied the difference in survival among people in different trajectory groups. Methods: Questionnaire data on work ability, working conditions, lifestyle, and physical functioning of middle-aged municipal employees (n = 2918) were linked with registers on retirement and all-cause mortality. Perceived work ability was measured as the current work ability compared with the lifetime best in a score of 0 to 10. The trajectory of work ability was analyzed by using growth mixture modeling in 16 years of follow-up data. Mobility limitations as an outcome was defined using nine items related to physical mobility tasks. Trajectory membership of work ability was used as a predictor of mobility limitation after 12 years using generalized linear models. Cumulative hazard curves for mortality by trajectory group were calculated. Results: Three distinct trajectories of work ability emerged. The majority of the participants (65%) had good work ability, which is here defined as sustainable work ability, 25% having L-shaped decreasing work ability and 10% having U-shaped decreasing work ability. Demographics, lifestyle factors, morbidity, and physical workload-adjusted models shows that L-shaped (Incidence rate ratio (IRR) 1.24, 95% confidence interval (CI) 1.18–1.30) and U-shaped (IRR 1.37, 95% CI 1.28–1.47) work ability trajectory membership was strongly associated with a higher risk of mobility limitations in the next 12 years of follow-up. The cumulative hazard for all-cause mortality was highest among those in the U-shaped decreasing work ability trajectory group. Conclusions: Those with a sustainable work ability during midlife showed a lower risk of mobility limitations and better survival compared to those with decreasing work ability. These findings highlight the importance of sustainable work ability throughout the working career as well as the need for early identification of workers with diminishing work ability and need for workplace interventions to help to promote an extended working career as well as a healthy old age.

Global Health Determinants and Limits to the Sustainability of Sustainable Development Goal 3

Eduardo Missoni

1. Introduction

The Millennium Declaration, signed by all Heads of State in September 2000, attempted to reaffirm a human-rights-based approach to development, with the liberation of the whole human race from need as one of its main goals (UN 2000). The Declaration proclaimed equality, freedom, solidarity, tolerance, respect for nature and shared responsibility as "fundamental values"; it also recognized the unequal distribution of common goods and the costs of globalization (UN 2000).

In order to translate the Declaration into a more operational instrument, it was later decided to identify the Millennium Development Goals (MDGs) that would make it possible to verify the progress of the development agenda. The original purpose of the MDGs was to go beyond the narrow paradigm of growth and focus on human, sustainable and equitable well-being. However, the conventional economic concept of development prevailed, with economic growth set as the primary force for poverty reduction (Vandemoortele 2010). Emphasis was placed on the identification of a limited number of relatively narrow targets and indicators, rather than on the need for deeper social transformations, almost entirely neglecting issues such as inequality and discrimination (Teichman 2014). The MDGs and the related targets set for 2015 lacked a systemic vision, and did not take the social, economic and environmental determinants of people's living and working conditions into account, or issues such as equity of distribution and access to resources (Fehling et al. 2013; Teichman 2014).

The MDGs focused exclusively on poor countries and reflected an idea of "development" pertaining only to development aid, involving high-income countries only as "donors". Indeed, low-income countries were scarcely involved in the process (Fehling et al. 2013).

In view of the 2015 deadline, in June 2012, the United Nations Conference on Sustainable Development "Rio +20", recorded a general consensus on the need for new global objectives for "concrete measures that accelerate implementation of sustainable development commitments" (UN 2012, p. 6), and initiated the inter-governmental process for the identification of new, truly global Sustainable Development Goals (SDGs). The result of that process was the Agenda 2030 for Sustainable Development, adopted on 25 September 2015 by the Summit of Heads of State and Government, convening in New York, by the United Nations (UN 2015).

In the traditional definition, sustainable development "meets the needs of current generations without compromising the ability of future generations to meet their own needs" (WCED 1987). Thus, sustainable development involves, on the one hand, the use of renewable resources and strict environmental protection, and on the other hand, the ability to ensure that human progress (first and foremost, the improvement of the living conditions of the populations) lasts over time.

The introduction to the ambitious Agenda 2030 affirms the "historic" dimension of the agreement, which commits governments to the adoption of a set of 17 "indivisible" objectives and 169 universal targets: to end poverty "once and for all" by 2030; to combat inequalities; to ensure lasting protection of the planet and its resources; to create the conditions for "shared prosperity" and "sustainable, inclusive and sustained" growth (UN 2015).

The new Agenda is not without contradictions. The achievement of "Sustainable, inclusive and sustained" economic growth, one of the pillars of the Agenda 2030 (UN 2015), is a conceptual oxymoron (Kopnina 2016; Spaiser et al. 2017).

Inside the planetary boundaries that define a safe operating space for humanity, "sustained" growth, with unmodified production and consumption patterns, is not compatible with sustainability. Almost fifty years ago, the first Report to the Club of Rome indicated the existing "Limits to growth" and called for "the initiation of new forms of thinking that will lead to a fundamental revision of human behaviour and, by implication, of the entire fabric of present-day society" to avoid "the tragic consequences of an overshoot" (Meadows et al. 1972, pp. 185–196). The forecast of a rapidly approaching global crisis based on mathematical models was recently confirmed, based on more solid data (Turner 2014). The current COVID-19 pandemic, and its global health, social and economic consequences, seems to be a dramatic expression of that forecast.

Nevertheless, the adoption of the "Agenda 2030 for Sustainable Development" (UN 2015) could still open a new phase in development policies in the global context of increasing complexity and unprecedented challenges, particularly as it stresses the interrelations between the different goals and their indivisibility.

Indeed, it was also suggested that health should be adopted as the main, if not the sole, goal of the sustainable development agenda, highlighting that health cross-cuts all phases of human life and people's individual and collective experience, such as education, work, gender balance, the distribution of wealth and access to resources, social protection, quality of the natural environment, capacity for self-determination and the quality of democracy. "Health is a dramatic and early indicator of the performance of other indicators, and equity in health measures the quality and extent of citizenship attributed to individuals in a society" (OISG 2014). In the same way, inequality in health is a mirror of all other inequalities, as well as constituting a "common danger", as stated in the WHO Constitution (WHO 2014).

Sustainable Development Goal 3 is to "ensure healthy lives and promote well-being for all at all ages". SDG 3 targets have been identified to measure the success of implemented policies and strategies. Among the nine health targets, the first three concern maternal mortality, infant mortality and the control of certain infectious diseases (primarily HIV/Aids, tuberculosis and malaria) and reflect previous MDGs. The new agenda adds targets for non-communicable diseases, substance abuse, deaths and injuries from road traffic accidents, universal access to reproductive health-care services, universal health coverage (UHC), and deaths and illnesses from hazardous chemicals and air, water and soil pollution and contamination. Finally, four specific actions have been identified, in particular (a) to strengthen the implementation of the "World Health Organization Framework Convention on Tobacco Control"; (b) to support the research and development of vaccines and medicines for developing countries, and ensure access to essential medicines and vaccines; (c) increase health financing, and the recruitment, development, training and retention of a health workforce in developing countries; (d) to strengthen the capacity of all countries for early warning, risk reduction and management of national and global health risks (UN 2015).

Although the challenge of sustainability is global, national health systems will be confronted with it based on the very different socio-economic conditions and expectations of their populations.

The achievement of UHC has been indicated as "the centrepiece of goal 3 … This is the one target that, if achieved—or let's say when achieved—will contribute to all the others" (Ghebreyesus 2018). UHC has been defined as universal access to the health services people need, when and where they need them, without financial hardship, including the full range of essential health services, from health promotion to prevention, treatment, rehabilitation, and palliative care (WHO 2020c).

It is generally accepted that universal access to quality care plays an important role in the improvement in population health and the reduction in health inequalities. In this sense, universal coverage is considered to be a particularly well-suited objective to address the complexity of the challenges facing health systems (Franklin 2017).

Among others, health systems' effectiveness and sustainability, i.e., the ability to take care of the needs of today without compromising the ability to provide for those needs in the future, is affected by a wider range of societal determinants beyond national boundaries.

Global power and processes may seriously undermine success, interfering with health systems' main functions, increasing demand or debilitating offers.

The acceleration of globalization and the hegemony of the neo-liberal ideology led to the progressive deregulation and liberalization of trade regimes, extensive privatization and the scaling back of the State. These processes have intensified the commodification and commercialization of vital social determinants such as health and social services, water and electricity. Unhealthy products are aggressively marketed by global industries (tobacco, alcohol, pesticides and other chemicals, processed foods and beverages, etc.). Environmental deterioration is also a result of the dominant economic model, which also heavily impacts labour and working conditions (CSDH 2008).

Health no longer depends solely on the specific situation of the country where people live but is largely determined by global forces acting outside the control of individual states, becoming an issue of foreign policy, global security, international trade, the overall sustainability of development, democratic governance and human rights (McInnes and Lee 2012).

Adopting Universal Health Coverage as a main focus, this chapter analyzes SDG 3's feasibility and sustainability using health system components as a framework to understand how global determinants interact with each component, affecting the functioning and sustainability of the health system as a whole. It concludes by arguing that, without an urgent paradigmatic shift in the current development model, the attainment of SDGs, and, specifically, the sustainability of SDG 3 and its "centerpiece" UHC will be at stake.

2. Health Systems' Functions

The World Health Organization (WHO) defined health systems as "all the activities whose primary purpose is to promote, restore and maintain health" (WHO 2000, p. 5). This includes health care as well as efforts to positively influence determinants of health. This conceptualization of health systems goes beyond the boundaries of the healthcare system, including policies and interventions often outside the direct competences of the health authorities, such as food accessibility, quality and safety, road safety or environmental control.

Based on the approach proposed by the WHO (2000; 2007; 2010b), the core objectives of health systems may be summarized as: (a) protecting and improving the health of the population they serve and reducing health inequalities; (b) responding to people's non-medical expectations and enabling participation in decisions that have an impact on their health and health systems; (c) protecting individuals from the risk of financial hardship due to the costs of health services through risk-pooling mechanisms, ensuring fairness for individual contributions and equity in access to services, i.e., access to and coverage of effective health interventions according to needs; (d) ensuring the best use of available resources to reach the aforementioned three objectives.

While the latter three objectives are specific to the healthcare system, the attainment of the first objective—protecting and improving health—relies only partially on healthcare and requires extending action, and even our understanding of a health system, to a system *for* health, i.e., beyond "activities whose *primary* purpose is to promote, restore and maintain health" (WHO 2000, p. 5)., to include all the policies and activities that have an impact on human health, thus also challenging the healthcare system, and its capacity to provide UHC, as will be discussed below.

Regardless of how they are organized, to achieve their goals, all health systems have to rely on some basic components: (a) leadership and governance; (b) human resources; (c) medical products and technology; (d) mobilization and allocation of finances; (e) service delivery. The components originally identified in the World Health Report 2000 were later summarized in the WHO Health Systems Framework, capturing information as an additional cross-cutting "building block" of increasing importance in supporting the overall functioning of the system (WHO 2007). The relationships among the six building blocks and their connection with the objectives of the health system are represented in Figure 1.

It is well-known that "health systems are subject to powerful forces and influences that often overwhelm the rational formulation of policies" (WHO 2010b). Among others, "these forces include a disproportionate focus on specialist care, fragmentation into a multiplicity of competing programs, projects and institutions, and the pervasive commercialization of health care into inadequately regulated systems" (WHO 2010b, p. 1).

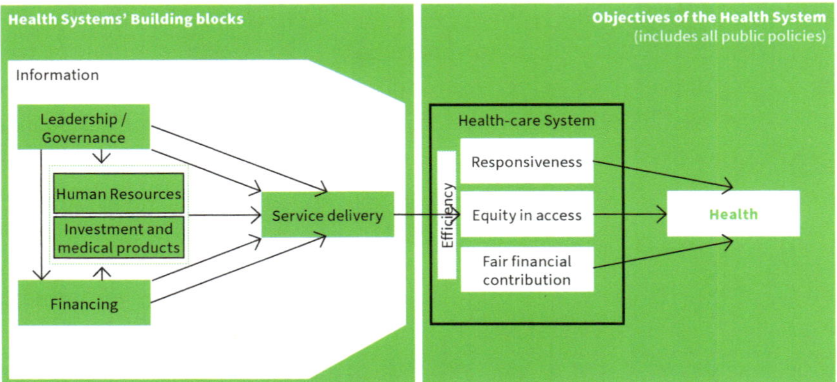

Figure 1. Health systems' building blocks and goals. Source: Modified from (WHO 2000; Missoni et al. 2019).

Indeed, the efficiency (appropriate use of resources) and effectiveness (achievement of objectives) and, ultimately, sustainability, of health care systems are put to the test by many forces and phenomena, which also require interventions and policies that are located outside the health system and often beyond the exclusive control of national authorities. In the following sections, we will focus on those interactions.

3. Determinants That Affect Steering and Governance

National health policies are influenced by international policies and transnational forces acting at different levels. Structurally or economically weaker states and economies are more susceptible to such influences and less prepared to deal with them. In the 1980s, international financial institutions (mainly the International Monetary Fund and the World Bank) imposed Structural Adjustment Plans (SAPs) on a large number of indebted countries, which entailed, among other things, a drastic reduction in public spending, the dismantling of universalist health systems, the privatization and commercialization of health services and the introduction of user fees—a real "tax on disease" (Geddes 2018, p. 35)—which had disastrous effects in terms of reduced access to services, exclusion of the weakest and the impoverishment of families. In more recent years, similar macroeconomic measures have been imposed by international and supranational bodies in more advanced economies as well, which were affected by the economic crisis, causing the impoverishment of large sections of the population (Kondilis et al. 2013). These measures contribute, among other things, to the fragmentation of health systems, making their management more complex and increasing their costs (Lister 2008; Geddes 2018).

In Official Development Assistance (ODA)-recipient countries, health policies and priority setting are strongly influenced by earmarked resources and donor conditionality, which often do not take the needs of partner countries into account (Biesma et al. 2009).

4. Determinants That Affect Human Resources and Access to Medical Products and Technology

Several global determinants influence the availability of human resources and access to medical products and technology.

The inadequacy of health workers' training in relation to the needs of the population has been recognized since the 1970s. The training of health personnel should not start from pathology, but from the context that generates disease, "from reality and not from theory, from the living society and not from the study of a corpse" (Maccacaro 1971, p. 377). With a few exceptions, medical faculties continue to follow what the Brazilian pedagogist Paulo Freire defined a "banking" educational approach (Torre et al. 2017), providing "information, or rather notions, detached from the context of real medicine that inevitably takes place more and more in the territory, outside the hospital" (Stefanini 2014). The World Health Report of 2008 also highlighted "hospital-centrism" among the problems at the root of the failure in achieving the health-for-all goal (WHO 2008). Practice in medical studies is mainly based on the observation of a hospitalized individual in a "horizontal" position, a "patient" in bed (Missoni 2018), and in a context too often socially and/or culturally alien to the social reality in which people "are born, live, work, grow old and die" (CSDH 2008, p. 26).

From the beginning of their career, future medical professionals are introduced to the logic of a globalized social model, which is profit-oriented and serves the interests of the dominant class (Stefanini 2014). Moreover, the standardization of skills and learning objectives (disproportionate focus on specialist curative care, high complexity, technological sophistication, etc.) respond to healthcare models that are scarcely sustainable even in middle–high income countries and are accessible elsewhere only to high-income population groups. Such an approach is a bad investment for low-income countries that already lack essential human resources. On the one hand, it produces health workers who are incapable of "usefully becoming part of an urban or rural community, of taking care of it, of understanding the problems of its illness and of defending its right to health" (Maccacaro 1971, pp. 377–382), with a training clearly detached from the local needs. On the other hand, the standardization of training across countries may have the unintended consequence of helping with professional migration across national boundaries (Frenk et al. 2010).

The main drivers of brain-drain include push (low level of income, poor working conditions, the absence of job openings and social recognition, oppressive political climate) and pull (better remuneration and working conditions, prospects for career development, job satisfaction, security) factors; however the latter leverages globally standardized training, which tends to produce "fit for export" health personnel. Indeed, health workers and, in particular, doctors, who are not prepared and unmotivated to serve in their own communities, will seek work elsewhere—first in large urban centers and then abroad—to obtain the kind of professional integration that requires their skills and meets the aspirations suggested in their medical studies, according to the globalized stereotype of the successful doctor (Missoni 2018).

Health workforce "brain drain", is also fueled by "import" agencies from high-income countries lacking human resources, often bypassing the norms that some of those countries have adopted based on the WHO global code (WHO 2010a). Healthcare-related trade policies and agreements have also promoted the migration of health professionals from the public sector to the private sector, and abroad.

Medical products and technologies are key resources for the health systems, but global determinants may affect their availability and affordability, as the following examples illustrate.

While basic research is still largely generated in universities and public research institutes, thanks to private funding, the research and development of biomedical products and devices are essentially in the hands of the transnational corporate sector, which invests only if a return on investment can be predicted, without taking health needs and health burden into account. As a result, only 10% of global health research spending is allocated to health conditions that account for 90% of the disease burden. Between 2000 and 2011, only 1% of the new active ingredients on the market were for neglected diseases (Pedrique et al. 2013). Thanks to the contribution of international initiatives and product development partnerships, there has been some progress in recent decades, especially for malaria, but very limited or no progress for other neglected diseases, such as dengue fever, Buruli ulcer, trachoma, rheumatic fever, or typhoid fever (Cohen et al. 2010).

The price of medicine remains the largest obstacle to access to care. Drugs and other medical products represent the largest public expenditure on health after personnel costs in many low-income countries, and the expense is a major cause of household impoverishment and debt (WHO 2020b).

Numerous patent-protected lifesaving medicines of proven efficacy are marketed at a high price, unaffordable for most of the people and healthcare systems in low-income countries. The global system of protection of intellectual property rights (IPR) may contribute to price increases and reduced access to medicines and vaccines (Smith et al. 2009). In the interest of public health, flexibilities under the Trade-Related Aspects of Intellectual Property Rights (TRIPS) agreement—such as compulsory licensing and parallel import—allow countries to gain access to medicines that, in other countries, may still be under patent. Unfortunately, there is still reluctance to implement these flexibilities due to concerns about the reactions of trade partners or a lack of the administrative, legal and/or productive capacity to adopt such an approach, which is also opposed by transnational pharmaceutical companies (Kerry and Lee 2007).

Prices may come down when the patent expires, and competition and/or generic products emerge. However, at that point, the transnational pharmaceutical industry often adopts sophisticated "lifecycle management" strategies. This patent "evergreening", based on the introduction of minor changes to the formulation, allows companies to extend their monopoly privileges on the drug, keep prices high and remain in control of these drugs on the market. This seriously challenges the access to affordable drugs, as it delays the generic competition without any improvement in the efficacy of the already-patented drug (Abbas 2019; WHO 2020b).

5. Determinants Affecting the Financing of the System

There is a correlation between the increase in health spending and increased life expectancy. However, above a spending threshold of about 75 US dollars per capita, that relationship becomes unpredictable and improvements in health outcomes depend mainly on the efficiency of the system (how money is spent) and on political choices related to social solidarity and equity (Savedoff et al. 2012).

In the same way, beyond a certain threshold of GDP per capita, economic growth is no longer correlated with health outcomes; rather, the inequality in the distribution of income directly correlates with disease burden (Pickett and Wilkinson 2015).

The efficiency of health care systems largely depends on how funds are collected, allocated, pooled and finally used for the purchase of services.

The combination of these components determines how many resources will be made available and how efficiently they will be used to achieve the desired results.

The macroeconomic framework of a country (including the quality and effectiveness of its fiscal system) determines resources' availability. Weaker economies have greater difficulty in bearing the costs of their health care systems and suffer most from financial crises and the imposition of international provisions (Gurtner 2010).

Ideologically mandated "rigorous" one-fits-all austerity policies impose social expenditure "cuts", including to salaries, maintenance costs and investments. As Geddes observes, "the undeclared objective" is to reduce the public services supported through the State's tax revenues, with the purely political aim of promoting the privatization of services, to the benefit of private capital (Geddes 2018), in a veritable "assault on universalism" (McKee and Stuckler 2011).

Dependence on out-of-pocket payments (OOPs) of services, introduce regressive mechanisms in financing, constitute a barrier to access to needed care and generate problems of financial protection. There is a very strong correlation between the level of OOPs and the incidence of catastrophic and impoverishing health expenditures, which are solely determined by the extent to which out-of-pocket payments absorb a household's financial resources (WHO 2020a).

Every year, more than 100 million people end up in poverty as a result of direct spending on health care (Haider and Nibb 2017). When in need, people without guarantees of access to care tend to turn to the much more expensive hospital emergency services. In addition, they tend to be excluded from health promotion and prevention activities that health services carry out.

In poor resource-settings, where health care providers tend to be inadequately paid, user fees may constitute a major source of revenue for health workers, creating perverse financial incentives (WHO 2020a).

Out-of-pocket health spending as a percentage of total health spending is highest in lower-middle-income countries and tends to make up a smaller share of the total health spending in upper-middle- and high-income countries (IHME 2019). However, even in the European Union "too many patients […] are facing financial hardships as a result of healthcare costs" (Franklin 2017, p. 2) and, for some people, direct costs act as a deterrent to the use of health services or to the continuity of care they cannot afford. Others, in order to access needed care, have to cut spending on food, clothing and housing. Unfortunately, since 2009, spending on direct payments in Europe has grown faster than public spending, with a negative impact on the functioning of health systems and the functioning of society in general, in terms of social cohesion and economic development (Franklin 2017).

Fragmentation reduces health care systems' efficiency, and systems based on the private coverage and insurance market are the most inefficient, with the sharpest increase in both public and private health expenditure (Unger and De Paepe 2019; Geddes 2018), which obviously makes this approach the less sustainable one.

Private insurance systems also promote over-diagnosis and induce health consumerism, without benefits to the public sector even in terms of waiting lists' reduction, use of services or cost reductions. Rather, cost-control is much more problematic in systems where different providers compete, even disregarding the negative impact on the equity of the system (Geddes 2018; Steendam et al. 2019).

In low-income countries, health services are largely dependent on development assistance. Unfortunately, external financing is often also volatile and unpredictable, making planning impossible. In addition, these funds are often tied to particular activities or diseases, sometimes do not respond to local needs and nationally established priorities and use autonomous management procedures and mechanisms with unnecessary duplication and increased transaction costs, generating an unsustainable administrative burden on already weak, resource-poor institutions (Missoni et al. 2019). The steady increase in the number of public and private aid actors and their profound diversity in terms of strategies and procedures, disregarding internationally agreed alignment and harmonization strategies (OECD 2005), further contribute to making aid inefficient and ineffective. Besides leading to an extreme and unsustainable fragmentation of national systems, the growing number of new private actors and public–private partnerships, pushed by the globalization of neoliberal ideology, also undermine the WHO's authority in directing and coordinating international health activities, and thus its leading role in support of national systems (Missoni et al. 2019; Ruger 2018).

6. Determinants That Affect Service Delivery

Both increases in demand and insufficient or inappropriate offers may affect the capacity of a healthcare system to provide universal access to care, and the sustainability of such a policy.

6.1. Demand

Besides demographic factors, the *fil rouge* that links multiple determinants of the unsustainable increase in illness and, as a consequence, demand is the consumerism inherent to the globalized capitalistic growth society, which is reflected in the multiple, diverse, but interconnected pathways described below.

The steady increase in the world's population and its progressive ageing are among the main causes of increased demand for health services. Between 2015 and 2050, the proportion of the world's population over 60 years of age will almost double, from 12% to 22% (WHO 2018b).

Ageing not only leads to a decline at the biological level, it is, in fact, associated with a profound transformation of a person's lifestyle, and often living conditions, which, in turn, may contribute to the worsening of physical and mental health. Although partly dependent on genetic factors, the health of elderly people is heavily influenced by social, economic and environmental determinants, including the quality of food, housing conditions and the consistency of family and community networks, as well as life experiences since early childhood. In this sense, the social determinants that affect young people today will influence the type and frequency of diseases in the coming decades. Geriatric syndromes that characterize health conditions of the elderly, i.e., complex multimorbidity, lead to a greater demand for health care and require totally new care approaches (WHO 2018b).

The considerable increase in the global burden from chronic diseases cannot be attributed exclusively to the ageing of the population. In fact, it affects all age groups and almost all countries, with a much greater impact in poorer countries, which are experiencing an epidemiological transition with a double burden of disease, i.e., both infectious diseases and chronic non-communicable diseases. Three quarters of deaths from chronic diseases are recorded in low- and middle-income countries (Haider and Nibb 2017).

While pandemics of old and new infectious diseases are seriously challenging humanity, an "epidemic" of chronic diseases, especially heart disease and cancer, observed since Second World War, clearly parallels the globalization of the western societal and lifestyle model (Kesteloot 2004).

A society aiming at and measuring its success through sustained growth requires a constant increase in consumption, no matter what is consumed and no matter the impact on the environment and health, based on the conventional wisdom that those are the unavoidable consequences of economic development (Landrigan et al. 2018). To that end, agricultural and industrial production cycles become faster, more resource-intensive and more contaminating, with the inexorable impoverishment of natural resources and increased levels of pollution, which is the largest environmental cause of disease and premature death in the world today (Landrigan et al. 2018).

With the promise of high yields and reduced losses in the production, pesticides and chemical fertilizers are massively promoted and used almost without control, causing contamination of the soil, water, and air, with a direct health hazard for rural workers and their families, and dangerous amounts of chemical residuals entering the food-chain, including drinking water (Kennedy et al. 2004; Landrigan et al. 2018;

Willett et al. 2019). The phases of food processing, packaging, transportation, and storage are also significant contributors to food contamination (Rather et al. 2017).

Similarly, industrial production is still widely based on the use of energy from fossil resources, contributing to high emissions of CO_2 and other greenhouse gases, and thus climatic change, as well as, in many cases, diffusion in the environment of many other dangerous contaminants, with a direct impact on the health of the population and an increase in health care expenditure in neighboring provinces (Zeng and He 2019). Climate change, indissolubly linked to the dominant production and consumption model, has an impact on health, with multiple interactions and predictable, potentially catastrophic, irreversible epidemiological transformations (Watts et al. 2017).

Extremely aggressive market strategies further push consumption. Worldwide, supermarkets' shelves are full of harmful food (processed foods with added sugar, salt, preservatives and colorants; high-calorie drinks; etc.), alcohol and tobacco, and other unhealthy or otherwise potentially harmful consumer products (such as home and personal care), which all contribute to the dramatic increase in chronic diseases such as obesity, metabolic diseases (first of all diabetes), respiratory diseases, cardiovascular, neoplastic, and neurodegenerative and mental illnesses (Kennedy et al. 2004; Landrigan et al. 2018; Willett et al. 2019).

In turn, consumption produces waste and a variety of pollutants, which, in most cases, are not optimally disposed of. Among others, the presence of microplastics in the food-chain, including treated tap- and bottled water, have raised considerable concerns regarding their impact on human health (WHO 2019).

Overall, environmental factors account for between 25% and 33% of the global burden of disease. A total of 83% of deaths are mediated by environmental factors. Carcinogenetic chemicals can now be found at every level of the food chain, in soils, groundwater and air emissions, and are widespread in a myriad of household and personal care products to which people are exposed every day (Haider and Nibb 2017).

The impact of pollutants on health is not limited to current generations. Besides the environmental consequences of the unsustainable exploitation of natural resources and the pollution of land, water and the atmosphere, many widely disseminated pollutants have been shown to produce epigenetic changes which are transmitted from one generation to the other, putting the health of future generations at risk (Skinner et al. 2010).

The ongoing commodification of water is the subject of growing concern in relation to water security, as well as quality- and water-related diseases, besides the maintenance of an unequitable status quo of inaccessibility. (Brisman et al. 2018).

Mental and relational pathologies (depression, suicide), and physical pathologies deriving from the use of new technologies (reduced physical exercise, pathologies of postural origin) are associated with globalized changes in lifestyle, including "behavioural addictions" associated with "excessive use of the internet, computers, smartphones and other electronic devices" which are also increasing (WHO 2015, p. 5).

The direct and indirect impact of the ever-increasing global exposure to electromagnetic fields on human health is widely underestimated and is a matter of increasing concern, calling for the adoption of severe precautionary principles (Bortkiewicz 2019).

Besides the pathological effects arising from the use of new technologies, the global expansion of the Internet has an additional impact on the increase in health demand. Social networks represent an easily accessible market of hundreds of millions of users through direct-to-consumer advertising of the improper or illegal use of often counterfeit medicines, with considerable health risks and an inevitable increase in health expenditure. Direct advertising is the fastest growing form of pharmaceutical marketing. Although only legal in the United States and New Zealand, online forms of interaction now allow legal restrictions to be violated everywhere (Liang and Mackey 2011).

Through disease-mongering strategies, i.e., creating patients, offering a distorted perception of the severity of a condition or presenting a physiological condition as pathological, the pharma industry induces the unnecessary consumption of drugs, contributing to the increase in health expenditure (Doran and Henry 2008).

The health care system itself also contributes to its own unsustainability, being one of the causes of the spread of antibiotic resistance, although 80% of antibiotic consumption happens in the livestock industry. The General Assembly of the United Nations, in 2016, warned against the potential re-emergence of diseases kept under control for decades, and the risk of new catastrophic epidemics (IACG 2019).

6.2. The Offer

Increased demand may also come from within the health care system. It is well known that, in the health sector, the increase in supply generates demand, particularly in the absence of control mechanisms and in health systems mainly based on private care. Particularly in developing countries, health care systems are highly fragmented and governments, which are mostly only in control of the public sector, are not able to create appropriate mechanisms to regulate the private sector's activities and performance. Experience in the Americas shows that fragmentation leads to difficulties in access to services, the poorer technical quality of services, the irrational and inefficient use of resources, unnecessary increases in production costs, and low user satisfaction with the services received (Montenegro et al. 2011).

On the other hand, health management in developing countries is inspired by the theories and practices adopted in high-income countries, and tends to reflect elements which are intimately linked to the technological, institutional and cultural characteristics of those countries. Management and governance systems are often imposed from above and are not consistent with the local context, while the "western" model, dominated by neoliberal policies, has become the universally adopted standard. In addition, we have been witnessing, for some time, a generalized attempt to transfer the logic, culture and managerial tools of private enterprises to the public sector, based on the unproven principle that market forces tend to generate better results than bureaucratic and hierarchical mechanisms (Fattore and Tediosi 2011).

Health, and health systems, is increasingly at the mercy of market dynamics. The commercialization of health care further contributes to the fragmentation of health systems, with resources being taken away from the public system to the advantage of the private system, with obvious discrimination in terms of the access to and quality of services, and even the exclusion of important sections of the population from access to both curative and preventive care (UNRISD 2007).

The lack of, or limited access to, adequate health care services (sometimes with paradoxical situations where family income is insufficient to afford private insurance, and not low enough to receive public assistance), pushes population groups with sufficient economic resources to travel for the purpose of receiving medical care at lower costs, engaging in so-called "Medical Tourism". While medical tourism is seen as an opportunity for economic returns in the recipient countries, eventually, many authors would agree that, for those countries, health tourism is a source of increased inequalities, possible overall healthcare cost increases and an additional push-factor for the migration of the health workforce from the public to the private sector, even without considering the risks for patient-tourists (related to travel, medical–surgical intervention, the course and post-operative care) (Hopkins et al. 2010).

It is paradoxical, and ethically unacceptable, to consider medical tourism as an opportunity that should be encouraged to reduce the national health expenditure of the originating countries, arguing, for example, that "If only 10 percent of the top

50 low-risk treatments were performed abroad, the U.S. health care system would save about \$1.4 billion annually" (Herrick 2007, p. 28), without worrying about the negative effects on the destination health care systems.

While technological innovation can contribute, among others, to promoting healthy behaviour, supporting home care, and facilitating more accurate diagnoses and better therapeutic responses, it is not always real progress and can create sustainability problems. Industry introduces new technologies responding to expectations of return on an investment, independently from the real therapeutic advantage they may offer (Thimbleby 2013). The health sector is also often prey to "planned obsolescence" as a market strategy by which manufacturers induce the replacement of equipment with new models that bring nothing substantive in terms of diagnostic or therapeutic results, aiming instead to create dependence on accessories and consumables (Rosenthal 2014). Most innovations tend to increase care costs, rather than reduce them, without a parallel increase in performance (Geddes 2018). What is too-often lacking is good management of existing technology and an adequate maintenance culture, an often-forgotten aspect in infrastructural and technological aid projects in low-income countries.

Over-prescription is another cause of increases in health care costs. Geddes (2018), for example, warns against "periodic check-ups", which are often promoted as part of well-designed market strategies in the biomedical industry, but have "no effect in reducing diseases and deaths from either cancer or cardiovascular disease", rather leading to an increase in diagnoses and "incidentalomas", with consequent risks related to further investigation.

The abuse of medicine, technologies and services, including ineffective or inappropriate use, is also linked to the culture and choices of prescribers (often under the marketing pressure of manufacturers and pharmaceutical representatives), patient requests (induced by misleading and increasingly pervasive advertising), conflicts of interest, fragmentation of levels of care leading to the repetition of clinical investigation, and remuneration criteria for facilities and professionals (Geddes 2018).

7. Making Health Systems Sustainable

The incompatibility between the planet's finite space and resources and the consumerist imperative of our global growth-society, as well as the impact on the health of the latter and the need for "degrowth", have been widely described (Missoni 2015).

Despite its contradictions (such as the oxymoron of "sustained sustainable growth"), which should be corrected, the implementation of Agenda 2030 and its universal sustainability goals represents an opportunity to avoid disaster. Indeed, pursuing "health and well-being for all and for all ages" (SDG 3) could make a fundamental contribution to the achievement of several other SDGs, including that of economic growth (Kieny et al. 2017). However, health-for-all in general, and the achievement of "universal access to essential quality healthcare services" in particular, are subject to the achievement of many other SDGs, where progress seems to be halting (UN 2019).

Despite the extraordinary international commitment, the necessary change in direction cannot yet be perceived. In the previous sections, we have tried to highlight how global determinants interact with the different components of national health systems, interfering with their sustainability and the achievement of the UHC goal. Below, some corrective actions are identified to promote the resilience of the system.

7.1. Steering and Governance

Intervention in the global determinants that act on the governance of the health system requires a solid alliance between public stakeholders, through all levels of the system, from the local (civil society organizations and grassroot movements), to the national and supranational (e.g., the EU), to the global (international institutions, in particular the WHO), both for the promotion of good practice and the prioritization of health in all public policies (economic, industrial, agricultural, social, etc.), when needed through regulatory interventions to control market forces that push in the opposite direction (Missoni 2015).

To this end, the WHO should be re-empowered and use all its authority to push public health needs and priorities in the political agenda of other sectors and initiatives, e.g., in trade negotiations, at both global and regional levels (Missoni 2015). The systematic application of the Health Impact Assessment (HIA) (WHO 1999) to all public policies would allow for the timely correction of legislative projects, plans and programs that do not respect that priority and may have negative health outcomes. Today, HIAs are implemented mostly as a section of the environmental, social and health impact assessment (ESHIA) of industrial projects (e.g., in the extractive industry), and are mostly implemented to meet a regulatory or statutory requirement. Instead, policy proposals should be subject to HIA with a wider societal scope, adopting a social view of health approaches and being used as a mechanism to address the potential health inequities that they may generate (Harris-Roxas and Harris 2011).

7.2. Ensuring the Right Human Resources and Access to Medical Products and Technology

It is essential to radically rethink health workers' curricula. They should be exposed to community health from the very beginning of their studies, and made aware of the role of socioeconomic and related social factors in shaping health. Their knowledge and experience should be consistently linked and relevant to local realities, as well as inspired by values of social solidarity and service, with people at the centre. Health professionals should be trained in the use of socially, culturally and economically appropriate technologies, and be guided by the needs of the population and the early correction of determinants with a negative impact on health.

Applying the WHO Code for the International Recruitment of Health Personnel (WHO 2010a) to address the shortage of health workers and their distribution will not be enough in the absence of an in-depth review of policies and investments aimed at aligning health workers' competences (knowledge, experience, motivation, values) to the context and needs of the population they are intended to serve, rather than investing in, and even increasing resources for, the perpetuation of a flawed system.

Similarly, it will be necessary to find the right incentives for the research and development of technologies to serve the health of the population, rather than the economic interests of investors. The market follows the consumer, and "corporate social responsibility" also responds, to a large extent, to consumers' choices. In the provision of health care, the main customer is often the State. In this case, the choices that decision-makers at different levels of the health system will be able to make in order to orient and regulate medical consumption and innovation will be particularly relevant.

7.3. Financing

Although there is no single formula for universal coverage, ensuring fairness in individual contributions and equity in access to services are essential to protect individuals from the risk of financial hardship due to the costs of health services.

To ensure fairness, individual contributions should be based on progressivity, i.e., increasing in rate as the base increases with the economic capacity of the contributor. This is a common characteristic of many fiscal systems, thus of health care systems that are financed through general taxation, an approach that many countries with large sectors of informal economy and/or weak fiscal systems cannot adopt, meaning that they need to recur to alternative mechanisms and sources to finance their public health care systems.

Besides a stable increase in the resources allocated to the health system, universal access to care requires greater use of advance payment mechanisms and a parallel reduction in the dependence on out-of-pocket payments for services, which are typically regressive and thus unfair (Haider and Nibb 2017). To increase efficiency, reduce inequalities and promote equity, it will be mandatory to reduce systems' fragmentation, merging collecting institutions, and, above all, pooling risk as widely as possible across the population.

The quality of financial and administrative management systems is also a further element in this direction (Evans and Antunes 2011). Stepping away from healthcare privatization policies is also essential to reduce inequality and costs, increase quality, efficiency and public control, and optimize the use of the health workforce (Steendam et al. 2019).

Certainly, to be sustainable, health systems need to be guided by long-term strategies that include investment in the development of adequate managerial approaches, simplification of processes, the appropriate use of ICT and a strengthened administrative capacity, in order to free up energy and human resources for care functions, reduce administrative complexity, and ensure proper monitoring of activities and costs and a reduction in waste (Geddes 2018). In this sense, integrated national health systems allow for considerable economies of scale and greater efficiency, in this as well as in terms of planning, procedural and technological standardization, and the centralization of procurement (Montenegro et al. 2011).

Given the globalization of health industries (pharmaceutical, insurance, etc.), global regulatory mechanisms and coordination are equally essential (Geddes 2018).

In countries where external aid plays a significant role, it is fundamental that funding, in addition to being increased, becomes more predictable (with long-term commitment), less fragmented and respectful of the well-known principles of aid effectiveness (ownership, alignment and harmonization) (OECD 2005). When feasible, General Budget Support initiatives and Sector-Wide Approach programs with the establishment of joint funds in support of sound national health plans are among the best options (Missoni et al. 2019).

7.4. Health Services

The response of the health care system to demographic changes (population growth and ageing) can only be through adaptation. The increased demand deriving from the ageing of the population, with its corollary of chronic and multi-morbidity diseases, calls for a rethinking of the model of care, and even of socialization, which can be implemented in the short- to medium-term. The active social integration of disabled and elderly people should be promoted as much as possible. In more advanced countries, experiences which are alternative to the conventional social organization (e.g., unifamilial home), such as extended families, life-communities, the sharing of living spaces (co-housing) and resources, offer opportunities to reconsider intergenerational experiences of solidarity, which are valid alternatives to the hospitalization and institutionalization of people with reduced autonomy (Missoni 2015). Naturally, such an approach also requires an investment in the development of new skills.

People and communities, with their own specific needs and problems, should be the focus of any system whose goal is health and wellbeing. A people-centred approach implies framing access to health care as a right and requires changes in both culture and society.

Universal access can only be guaranteed through a Primary Health Care approach, which was already indicated in 1978 as the strategy to ensure "Health for all by the year 2000", but which was immediately opposed and replaced by the selective approach that has contributed everywhere to weakened health systems (Missoni et al. 2019). Primary care must link the community and the rest of the health care system, i.e., it must be the basic element of a care system, which links social protection and support to health care and is organized according to well-connected levels of complexity and intensity of care that are able to provide appropriate responses at the most appropriate level. Pathways of integration among different policy fields, including health, social protection and urban regeneration, have been promoted and encouraged to address societal changes, particularly in more deprived urban areas (De Vidovich 2020). Linking the health care system to community social networks (volunteering, self-help groups, self-managed centers, etc.) and involving the patient as an active player in the care processes is also part of the Chronic Care Model (CCM), and its further development also responds to this need (Wagner et al. 2001).

In many countries, primary care is a natural hub for the integration of Traditional and Complementary Medicine (T&CM) with national health systems. This integration has been advocated since the 1970s to improve primary care access and health outcomes by increasing the availability of services as an additional point of contact and a clear contribution to pathways toward UHC (Lee Park and Canaway 2019). However, in many cases, a lack of support or frank opposition from central governments, institutions, medical organizations, and the biomedical industry represented an obstacle in that direction. The considerable worldwide use of T&CM, both in industrialized countries (Unites States, 42%; Australia, 48%; France, 49%; Canada, 70%) and less advanced economies (Chile, 71%; Colombia, 40%; up to 80% in African countries), should suggest the need to bring traditional medicine "into the mainstream of health care, appropriately, effectively, and above all, safely" (WHO 2018a, p. 1).

Today the connection between primary care and a higher level of complexity may take advantage of new information and communication technologies (e-health, m-health, big-data, social networks, etc.); however, these also require guidance, regulation and organization within the health system for their optimization. The digital revolution of the health system is, in fact, a "Tsunami" directed by "numerous, powerful and intelligent forces and actors" with an "immense thirst for technological and economic conquest" (Comtesse 2017). Health systems are not prepared to face this challenge. Due to its transnational dimension, the response needs a phenomenal commitment to global analysis and direction. In general, a systemic approach to new technologies is still missing almost everywhere, and the projects described in the literature refer almost exclusively to pilot experiences that seem to lack systematization. The WHO itself states that: "For eHealth to play its full role in helping health systems achieve UHC a sound legal framework is required", which is obviously still missing (WHO, p. 6).

In general, the focus must be shifted from treatment to primary prevention, from the hospital to the community—now also the virtual one—where the disease originates. Therefore, first of all, policies and interventions are needed aimed at improving the daily living and working conditions of the population (housing, workplace, public spaces, transport, recreational and sports facilities, etc.), as well as policies aiming to "tackle the inequitable distribution of power money and resources" (CSDH 2008, pp. 108–109). Interventions at the community level involve local and national responsibilities and public policies, which should prioritize health and inevitably involve multiple sectors and authorities outside the health sector. Nevertheless, the role of the choices made at the international level to promote and support national and local initiatives should not be underestimated.

The control of internet-mediated activities on health and health consumption necessarily requires international synergies, as national laws are easily circumvented on the web, making global measures urgent (Liang and Mackey 2011).

The primary (i.e., immunization) and secondary prevention (i.e., screening and early detection) programs offered by health services are of great importance. However, policy decisions should always consider risks in terms of safety, effectiveness and possible risks.

In the long-term, the reduction in demand due to chronic diseases can only derive from interventions for systemic determinants, for example, through laws and regulations that impose (e.g., through restrictions, taxes and other disincentives) the internalization of the social costs of production cycles, practices, services and products that are harmful to health, which are otherwise transferred to the community, while incentivizing the accessibility (wide availability, lower costs, etc.) of healthy products and services. Unfortunately, this type of intervention clashes with strong interests and requires considerable courage and political will. Health education campaigns aimed at promoting the change in individual behaviors (stopping smoking, reducing the consumption of alcohol or sugar, doing more exercise, consuming healthy food, etc.) are an easier alternative (politically and socially less problematic), but considerably less effective (Swinburn et al. 2011). At the local level, changes in consumption patterns and behavior can greatly benefit from the initiatives of single groups and communities. Instead, at a global level, only a strong connection, through national and transnational networks of civil society organizations, and their cooperation with international institutions, particularly the WHO by mandating "the directing and coordinating authority on international health work", will generate the alliance which is necessary to combat the "globalization of unhealthy lifestyles" and oppose "the commercial interests of powerful economic operators" (Chan 2013).

8. Conclusions

Despite good intentions, the achievement of SDGs seems to move further away every year: "At the current pace, around 500 million people could remain in extreme poverty by 2030. Global hunger is on the rise. Violent conflicts, climate change, gender disparities, and persistent inequalities are undermining efforts to achieve the SDGs" (Steiner 2019, p. 1). The COVID-19 pandemic is an additional, unprecedented wake-up call.

More than ever, the achievement of the universal and indivisible SDGs set by Agenda 2030 represents a considerable challenge for health systems worldwide.

UHC is a central goal for the health sector in the wider context of SDG 3. However, the feasibility and sustainability of universal access is heavily dependent on the intertwined actions of multiple and diverse forces and determinants acting at various levels, with global determinants playing an enormous role. Thus, pursuing "health and wellbeing for all at all ages" will require strong intersectoral collaboration and pushing the health priority into all public policies. The success of Agenda 2030 undoubtedly rests on the respect of the principle of indivisibility of the SDGs.

From the examples proposed in this paper, it should be evident that the main "cause of causes" of the unsustainability of the UHC target, and of SDG 3 as a whole, is the dominant neoliberal, market-oriented societal model. As a consequence of its indivisibility, the same overarching determinant hampers the success of the whole Agenda 2030.

The scientists who wrote the report commissioned by the Club of Rome were "convinced that realization of the quantitative restraints of the world environment and of the tragic consequences of an overshoot is essential to the initiation of new forms of thinking that will lead to a fundamental revision of human behavior and, by implication, of the entire fabric of present-day society" and called for a "basic change of values and goals at individual, national, and world levels" (Meadows et al. 1972, pp. 189–190). Today, evidence of the correctness of that analysis and forecast makes that radical economic and social transformation imperative (Turner 2014).

The (un)sustainability of the UHC target and SDG 3 may be a good indicator of the limits of the Agenda 2030 in the absence of a paradigmatic shift toward a more inclusive, cooperative, equitable and ecological human society, where nobody is left behind.

Conflicts of Interest: The author declares no conflict of interest.

References

Abbas, Muhammad Z. 2019. Evergreening of pharmaceutical patents: A blithe disregard for the rationale of the patent system. *Journal of Generic Medicines* 15: 53–60. [CrossRef]

Biesma, Regien C., Ruairi Brugha, Andrew Harmer, Aisling Walsh, Neil Spicer, and Gill Walt. 2009. The effects of global health initiatives on country health systems: A review of the evidence from HIV/AIDS control. *Health Policy Plan* 24: 239–52. [CrossRef] [PubMed]

Bortkiewicz, Alicja. 2019. Health effects of Radiofrequency Electromagnetic Fields (RF EMF). *Industrial Health* 57: 403–405. [CrossRef] [PubMed]

Brisman, Avi, Bill McClanahan, Nigel South, and Reece Walters. 2018. *Water, Crime and Security in the Twenty-First Century Too Dirty, Too Little, Too Much.* Critical Criminological Perspectives. London: Palgrave McMillan.

Chan, Margaret. 2013. WHO Director-General Addresses Health Promotion Conference. Presented at the Opening Address at the 8th Global Conference on Health Promotion, Helsinki, Finland, June 10; Available online: http://www.who.int/dg/speeches/2013/health_promotion_20130610/en/ (accessed on 4 July 2020).

Cohen, Joshua, Maria S. Dibner, and Andrew Wilson. 2010. Development of and Access to Products for Neglected Diseases. *PLoS ONE* 5: e10610. [CrossRef] [PubMed]

Comtesse, Xavier. 2017. *Santé 4.0. Le Tsunami du numérique.* Chêne-Bourg: Georg Editeur.

CSDH. 2008. *Closing the Gap in a Generation Health Equity through Action on the Social Determinants of Health.* Final Report of the Commission on Social Determinants of Health. Geneva: World Health Organization.

De Vidovich, Lorenzo. 2020. Urban regeneration through the territorialisation of social policies: Findings from the Microareas Programme in Trieste, Italy. *Journal of Urban Regeneration & Renewal* 13: 406–21.

Doran, Evan, and David Henry. 2008. Disease mongering: Expanding the boundaries of treatable disease. *Internal Medicine Journal* 38: 858–861. [CrossRef]

Evans, David, and Adelio Antunes. 2011. Achieving universal health coverage: Supporting the development of national health financing systems. In *Attaining Universal Health Coverage. A Research Initiative to Support Evidence-Based Advocacy and Policy-Making.* Edited by Missoni Eduardo. Milano: Egea, pp. 15–29.

Fattore, Giovanni, and Fabrizio Tediosi. 2011. Attaining universal health coverage: The role of governance and management. In *Attaining Universal Health Coverage. A Research Initiative to Support Evidence-based Advocacy and Policy-Making.* Edited by Missoni Eduardo. Milano: Egea, pp. 31–51.

Fehling, Maya, Brett D. Nelson, and Sridhar Venkatapuram. 2013. Limitations of the Millennium Development Goals: A literature review. *Global Public Health* 8: 1109–22. [CrossRef]

Franklin, Paula. 2017. Sustainable Development Goal on Health (SDG3): The opportunity to make EU health a priority. *EPC Discussion Paper*, May 18.

Frenk, Julio, Lincoln Chen, Zulfiqar Bhutta, Jordan Cohen, Nigel Crisp, Timothy Evans, Harvey Fineberg, Patricia Garcia, Yang Ke, Patrick Kelley, and et al. 2010. Health professionals for a new century: Transforming education to strengthen health systems in an interdependent world. *The Lancet* 376: 1923–58. [CrossRef]

Geddes, Marco. 2018. *La Salute Sostenibile. Perché Possiamo Permetterci un Servizio Sanitario equo ed Efficace.* Roma: Il Pensiero Scientifico Editore.

Ghebreyesus, Tedros Adhanom. 2018. *Launch of Sustainable Development Goals Themed Issue of the Bulletin of the World Health Organization.* Geneva: World Health Organization, August 31, Available online: https://www.who.int/dg/speeches/2018/sustainable-development-goals-who-bulletin-launch/en/ (accessed on 4 July 2020).

Gurtner, Bruno. 2010. The Financial and Economic Crisis and Developing Countries. *International Development Policy | Revue Internationale de Politique de Développement* 1: 189–213. [CrossRef]

Haider, Muhiuddin, and Katrina Nibb. 2017. Universal Health Coverage and Environmental Health: An Investigation in Decreasing Communicable and Chronic Disease by Including Environmental Health in UHC. In *Advancement in Health Management.* Edited by Ubaldo Comite. Rijeka: IntechOpen, pp. 129–46.

Harris-Roxas, Ben, and Elizabeth Harris. 2011. Differing forms, differing purposes: A typology of health impact assessment. *Environmental Impact Assessment Review* 31: 396–403. [CrossRef]

Herrick, Devon M. 2007. *Medical Tourism: Global Competition in Health Care*. NCPA Policy Report N. 304. Dallas: National Center for Policy Analysis.

Hopkins, Laura, Ronald Labonté, Vivien Runnels, and Corinne Packer. 2010. Medical tourism today: What is the state of existing knowledge? *Journal of Public Health Policy* 31: 185–98. [CrossRef] [PubMed]

IACG. 2019. *No Time to Wait: Securing the Future from Drug-Resistant Infections*. Report to the Secretary-General of the United Nations. UN Interagency Coordination Group. New York: United Nations, April.

IHME. 2019. *Financing Global Health 2018: Countries and Programs in Transition*. Seattle: Institute for Health Metrics and Evaluation.

Kennedy, Gina, Guy Nantel, and Parkash Shetty. 2004. Globalization of food systems in developing countries: A synthesis of country case studies. In *Globalization of Food Systems in developing Countries: Impact on Food Security and Nutrition*. FAO—Food and Nutrition Paper 83. Rome: Food and Agriculture Organization, pp. 1–25.

Kerry, Vanessa B., and Kelley Lee. 2007. TRIPS, the Doha declaration and paragraph 6 decision: What are the remaining steps for protecting access to medicines? *Globalization and Health* 3: 1–12. [CrossRef]

Kesteloot, Hugo. 2004. Epidemiology: Past, present and future. *Verh K Acad Geneeskd Belg* 66: 384–405.

Kieny, Marie P., Henk Bekedam, Delanyo Dovlo, James Fitzgerald, Jarno Habicht, Graham Harrison, Hans Kluge, Vivian Lin, Natela Menabde, Zafar Mirza, and et al. 2017. Strengthening health systems for universal health coverage and sustainable development. *Bulletin of the World Health Organization* 95: 537–9. [CrossRef]

Kondilis, Elias, Stathis Giannakopoulos, Magda Gavana, Ioanna Ierodiakonou, Howard Waitzkin, and Alexis Benos. 2013. Economic crisis, restrictive policies, and the population's health and health care: The Greek case. *American Journal Public Health* 103: 973–9. [CrossRef]

Kopnina, Helen. 2016. The victims of unsustainability: A challenge to sustainable development goals. *International Journal of Sustainable Development & World Ecology* 23: 113–21.

Landrigan, Philip J., Richard Fuller, Nereus J. R. Acosta, Olusoji Adeyi, Robert Arnold, Niladri Basu, Abdoulaye Bibi Baldé, Roberto Bertollini, Stephan Bose-O'Reilly, Jo Ivey Boufford, and et al. 2018. The Lancet Commission on pollution and health. *The Lancet* 391: 462–512. [CrossRef]

Lee Park, Yu, and Rachel Canaway. 2019. Integrating Traditional and Complementary Medicine with National Healthcare Systems for Universal Health Coverage in Asia and the Western Pacific. *Health Systems & Reform* 5: 24–31.

Liang, Bryan A., and Timothy K. Mackey. 2011. Prevalence and Global Health Implications of Social Media in Direct-to-Consumer Drug Advertising. *Journal of Medical Internet Research* 13: e64. [CrossRef]

Lister, John. 2008. WHO Commission on Social Determinants of Health: Globalization and Health Systems Change. In *Globalization and Health Knowledge Network Research Papers*. Ottawa: Institute of Population Health.

Maccacaro, Giulio A. 1971. Una facoltà di medicina capovolta. Intervista. *Tempo Medico* 97: 377–382.

McInnes, Colin, and Kelley Lee. 2012. *Global Health and International Relations*. Cambridge: Polity.

McKee, Martin, and David Stuckler. 2011. The assault on universalism: How to destroy the welfare state. *BMJ* 343: d7973. [CrossRef] [PubMed]

Meadows, Donella H., Dennis L. Meadows, Jorgen Randers, and William W. Behrens III. 1972. *The Limits to Growth. A Report for the Club of Rome's Project on the Predicament of Mankind*. New York: Universe Books.

Missoni, Eduardo. 2015. Degrowth and health: Local action should be linked to global policies and governance for health. *Sustainability Science*. [CrossRef]

Missoni, Eduardo. 2018. Determinanti globali della salute e (in)sostenibilità dell'obiettivo di copertura sanitaria universale. *Welfare e Ergonomia* 1: 11–38. [CrossRef]

Missoni, Eduardo, Gugliemo Pacileo, and Fabrizio Tediosi. 2019. *Global Health Governance and Policy: An Introduction*. Abingdon: Routledge.

Montenegro, Hernán, Reynaldo Holder, Caroline Ramagem, Soledad Urrutia, Ricardo Fabrega, Renato Tasca, Gerardo Alfaro, Osvaldo Salgado, and Maria Angelica Gomes. 2011. Combating Health Care Fragmentation through Integrated Health Service Delivery Networks in the Americas: Lessons Learned. *Journal of Integrated Care* 19: 5–16. [CrossRef]

OECD. 2005. *Paris Declaration on Aid Effectiveness*. Paris: High Level Forum, 28 February–2 March.

OISG. 2014. La Salute al Centro dell'agenda Post 2015. Osservatorio Italiano sulla Salute Globale. Available online: http://www.saluteglobale.it/index.php/2014/09/02/la-salute-al-centro-dellagenda-post-2015/ (accessed on 4 July 2020).

Pedrique, Belen, Nathalie Strub-Wourgaft, Claudette Some, Piero Olliaro, Patrice Trouiller, Nathan Ford, Bernard Pécoul, and Jean-Hervé Bradol. 2013. The drug and vaccine landscape for neglected diseases (2000–2011): A systemic assessment. *Lancet Glob Health* 1: e371–9. [CrossRef]

Pickett, Kate, and Richard Wilkinson. 2015. Income inequality and health: A causal review. *Social Science & Medicine* 128: 316–26.

Rather, Irfan A., Wee Yin Koh, Woon K. Paek, and Jeongheui Lim. 2017. The Sources of Chemical Contaminants in Food and Their Health Implications. *Frontiers Pharmacology* 17: 830. [CrossRef]

Rosenthal, E. 2014. Even Small Medical Advances Can Mean Big Jumps in Bills. *The New York Times*. April 5. Available online: https://www.nytimes.com/2014/04/06/health/even-small-medical-advances-can-mean-big-jumps-in-bills.html (accessed on 4 July 2020).

Ruger, Jennifer Prah. 2018. *Global Health Justice and Governance*. Oxford: Oxford University Press.

Savedoff, William D., David de Ferranti, Amy L. Smith, and Victoria Fan. 2012. Political and economic aspects of the transition to universal health coverage. *The Lancet* 380: 924–32. [CrossRef]

Skinner, Michael K., Mohan Manikkam, and Carlos Guerrero-Bosagna. 2010. Epigenetic transgenerational actions of environmental factors in disease etiology. *Trends in Endocrinology & Metabolism* 21: 214–22.

Smith, Richard D., Carlos Correa, and Cecilia Oh. 2009. Trade TRIPS and pharmaceuticals. *The Lancet* 373: 684–91. [CrossRef]

Spaiser, Viktoria, Shyam Ranganathan, Ranjula Bali Swain, and David J. T. Sumpter. 2017. The sustainable development oxymoron: Quantifying and modelling the incompatibility of sustainable development goals. *International Journal of Sustainable Development & World Ecology* 24: 457–70.

Steendam, Julie, Chiara Bodini, and André Crespin. 2019. *Why Public Health Care Is Better*. Vivasalud: Brussels.

Stefanini, Aangelo. 2014. "Capovolgere" la facoltà di medicina? L'eredità di Giulio A. Maccacaro. *Salute Internazionale*. April 22. Available online: http://www.saluteinternazionale.info/2014/04/capovolgere-la-facolta-di-medicina-leredita-di-giulio-a-maccacaro/#biblio (accessed on 7 September 2019).

Steiner, Achim. 2019. Statement at the 100th Meeting of the Development Committee. October 19. Available online: http://documents1.worldbank.org/curated/en/145291572031527413/pdf/Statement-by-Achim-Steiner-at-the-100th-meeting-of-the-Development-Committee-held-on-October-19-2019.pdf (accessed on 25 March 2021).

Swinburn, Boyd A., Gary Sacks, Kevin D. Hall, Klim McPherson, Diane T. Finegood, Marjory L. Moodie, and Steven L. Gortmaker. 2011. The global obesity pandemic: Shaped by global drivers and local environments. *The Lancet* 378: 804–14. [CrossRef]

Teichman, Judith. 2014. *The MDGs and the Issue of Inequality: The Enduring Power of Neoliberalism*. CCDS Working Paper #1. Toronto: Centre for Critical Development Studies (CCDS).

Thimbleby, Harold. 2013. Technology and the future of healthcare. *Journal of Public Health Research* 2: e28. [CrossRef]

Torre, Dario, Victoria Groce, Richard Gunderman, John Kanter, Steven Durning, and Steven Kanter. 2017. Freire's view of a progressive and humanistic education: Implications for medical education. *MedEdPublish* 6: 5. [CrossRef]

Turner, Graham. 2014. *Is Global Collapse Imminent?* MSSI Research Paper No. 4. Melbourne: Melbourne Sustainable Society Institute, the University of Melbourne.

UN. 2000. *United Nations Millennium Declaration*. United Nations General Assembly, Fifty-Fifth Session. Agenda Item 60 (b). New York: United Nations.

UN. 2012. *Outcome of the Conference The Future We Want*. Rio de Janeiro, Brazil, 20–22 June, A/CONF.216/L.1. New York: United Nations.

UN. 2015. *Transforming Our World: The 2030 Agenda for Sustainable Development*. New York: United Nations, September 25.

UN. 2019. *The Sustainable Development Goals Report 2019*. New York: United Nations.

Unger, Jean-Pierre, and Pierre De Paepe. 2019. Commercial Health Care Financing: The Cause of U.S., Dutch, and Swiss Health Systems Inefficiency? *International Journal of Health Services* 49: 431–56. [CrossRef] [PubMed]

UNRISD. 2007. Commercialization and Globalization of Health Care: Lessons from UNRISD Research. UNRISD Research and Policy Brief 7. Available online: https://www.unrisd.org/80256B3C005BCCF9/httpNetITFramePDF?ReadForm&parentunid=7C04740AD2852A4AC12573E6002E24DC&parentdoctype=brief&netitpath=80256B3C005BCCF9/(httpAuxPages)/7C04740AD2852A4AC12573E6002E24DC/$file/RPB7e.pdf (accessed on 25 March 2021).

Vandemoortele, Jan. 2010. *Changing the Course of MDGs by Changing the Discourse*. Area International Relations and Development, ARI 132/2010. Madrid: Real Instituto Elcano, September 9.

Wagner, Edward H., Brian Anthony Austin, Connie Davis, Mike Hindmarsh, Judith Schaefer, and Amy Bonomi. 2001. Improving chronic illness care: Translating evidence into action. *Health Affairs* 20: 64–78. [CrossRef] [PubMed]

Watts, Nick, Markus Amann, Sonja Ayeb-Karlsson, Kristine Belesova, Timothy Bouley, Maxwell Boykoff, Peter Byass, Wenjia Cai, Diarmid Campbell-Lendrum, Jonathan Chambers, and et al. 2017. The Lancet countdown on health and climate change: Form 25 years of inaction to a global transformation for public health. *The Lancet* 391: 581–630. [CrossRef]

WCED. 1987. *Our Common Future. World Commission on Environment and Development*. Oxford: Oxford University Press.

WHO. 1999. *Health Impact Assessment: Main Concepts and Suggested Approach*. Gothenburg Consensus Paper. Brussels: European Centre for Health Policy, World Health Organization Regional Office for Europe.

WHO. 2000. *World Health Report 2000*. Geneva: World Health Organization.

WHO. 2007. *Everybody's Business—Strengthening Health Systems to Improve Health Outcomes—WHO's Framework for Action*. Geneva: Word Health Organization.

WHO. 2008. *World Health Report 2008. Primary Health Care, Now More Than Ever*. Geneva: World Health Organization.

WHO. 2010a. *The WHO global code of practice on the international recruitment of health personnel*. Geneva: World Health Organization.

WHO. 2010b. *Key Components of a Well Functioning Health System*. Geneva: World Health Organization.

WHO. 2014. Constitution of the World Health Organization. In *Basic Documents—48th ed*. Including Amendments Adopted up to 31 December 2014. Geneva: World Health Organization.

WHO. 2015. *Public Health Implications of Excessive Use of the Internet, Computers, Smartphones and Similar Electronic Devices: Meeting Report*. Tokyo: Main Meeting Hall, Foundation for Promotion of Cancer Research, National Cancer Research Centre, 27–29 August.

WHO. *Global Diffusion of eHealth: Making Universal Health Coverage Achievable. Report of the Third Global Survey on eHealth*. Geneva: World Health Organization.

WHO. 2018a. *Traditional and Complementary Medicine in Primary Health Care*. Technical Series on Primary Health Care, WHO/HIS/SDS/2018.37; Geneva: World Health Organization.

WHO. 2018b. Ageing and Health. Available online: http://www.who.int/en/news-room/fact-sheets/detail/ageing-and-health (accessed on 4 July 2019).

WHO. 2019. *Microplastics in Drinking-Water*. Geneva: World Health Organization.

WHO. 2020a. Out-of-Pocket Payments, User Fees and Catastrophic Expenditure. Available online: https://www.who.int/health_financing/topics/financial-protection/out-of-pocket-payments/en/ (accessed on 4 July 2020).

WHO. 2020b. Medicines. Available online: https://www.who.int/health-topics/medicines#tab=tab_1 (accessed on 4 July 2020).

WHO. 2020c. *Universal Health Coverage*. Geneva: World Health Organization, Available online: https://www.who.int/health-topics/universal-health-coverage#tab=tab_1 (accessed on 4 July 2020).

Willett, Walter, Johan Rockström, Brent Loken, Marco Springmann, Tim Lang, Sonja Vermeulen, Tara Garnett, David Tilman, Fabrice DeClerck, Amanda Wood, and et al. 2019. Food in the Anthropocene: The EAT–Lancet Commission on healthy diets from sustainable food systems. *The Lancet* 393: 447–92. [CrossRef]

Zeng, Juying, and Qiuqin He. 2019. Does industrial air pollution drive health care expenditures? Spatial evidence from China. *Journal of Cleaner Production* 218: 400–8. [CrossRef]

Transitioning to Good Health and Well-Being: The Essential Role of Breastfeeding

Colin Binns, Mi Kyung Lee and Jane Scott

1. Introduction

In the past few decades, there have been major improvements in the health of the citizens of most countries in our world. World infant mortality fell from 160 deaths under 12 months per 1000 live births in 1950, to 41 in 2016. There are still wide discrepancies between the regions with the highest infant mortality (West and Central Africa; 95 per 1000 live births) and the lowest (Western Europe; 4 per 1000 live births) rates (UNICEF 2017). There are also variations between countries and within the districts of countries. In the Sustainable Development Goals (SDG) 2018 progress report on SDG 3 *Ensure healthy lives and promote well-being for all at all ages*, the UN stated that, while the quality of global health is increasing, "people are still suffering needlessly from preventable diseases, both infectious and non-communicable, "and too many are dying prematurely" (United Nations Economic and Social Council 2019).

Promoting, and achieving, the World Health Organization (WHO) goals for breastfeeding is essential for achieving SDG 3. The WHO recommends "mothers worldwide to exclusively breastfeed infants for the child's first six months to achieve optimal growth, development and health. Thereafter, they should be given nutritious complementary foods and continue breastfeeding up to the age of two years or beyond" (World Health Organization 2011). These recommendations are consistent with the historical cultural practices of all societies, and the promotion of breastfeeding is a return to a pre-commercial era.

Breastmilk provides complete nutrition for an infant, supporting optimal growth and development, for the first six months of life, and protects against some childhood diseases and some chronic disease later in life (Horta and Victora 2013). Physiologically, almost all mothers can breastfeed, and sometimes grandmothers or relatives can re-lactate to feed an infant if a mother is prevented from doing so by illness. In some lower–middle income countries (LMICs), such as the Maldives and Papua New Guinea, most all mothers initiate breastfeeding with rates of almost 100% (Binns 1976; Raheem et al. 2014). Some Scandinavian countries have almost reached this level and, in Denmark, rates of initiation of 99% have been reported (Maastrup et al. 2019). With the help of family, the health system and society, and a supportive environment, high rates of breastfeeding are possible for all mothers. There are, of course, a few situations where breastfeeding is not possible (World Health Organization 2009). These situations include seriously ill, low birthweight infants and some maternal and infant infections. Many of these infants benefit from expressed breastmilk, or the feeding of breastmilk can be commenced later.

The aim of this review is to demonstrate the central role that the promotion of breastfeeding has to achieving the Sustainable Development Goals.

2. Materials and Methods

This review was based on a search of the English language literature using the keywords 'breastfeeding', 'sustainability' and 'infant feeding guidelines'. The Web of Science, Science Direct and PubMed databases were used with no time restrictions. All studies on the importance of breastfeeding are limited by the ethical restraints of undertaking research on infants in such a critical field (Binns et al. 2017). Most studies are, therefore, observational in nature, but the accumulation of evidence over many studies reassures of the benefits of breastfeeding for infants, their mothers and society.

3. Results: Sustainable Development Goal 3 Specific Targets

Specific targets have been set within each of the Sustainable Development Goals, including specific health targets for SDG 3 (World Health Organization 2017). The promotion of breastfeeding will contribute to most of these targets, in conjunction with other health interventions.

3.1. Reduce Global Maternal Mortality

By 2030, reduce the global maternal mortality ratio to less than 70 per 100,000 live births (World Health Organization 2017).

The global maternal mortality rate has fallen in recent years, from an estimated 451,000 maternal deaths in 2000 to 295,000 in 2017. The maternal mortality ratio is now 211 maternal deaths per 100,000 live births (World Health Organization 2019a). This is a continuing public health tragedy that means that an estimated 808 mothers are dying in childbirth every day. The global lifetime risk of maternal mortality for a 15-year-old girl in 2017 was estimated at 1 in 190. During the period 2012–2017, almost 80% of live births worldwide occurred with the assistance of skilled health personnel, up from 62% in 2000 to 2005 (United Nations

2019). Early initiation of breastfeeding, as advocated in the Baby Friendly Hospital Initiative, raises oxytocin levels after delivery, causing the contraction of the uterus, and may reduce post-partum hemorrhage (Sentilhes et al. 2016; WHO UNICEF 2018). This would have a particular benefit in LMICs of reducing post-partum hemorrhage, which continues to be an important cause of maternal deaths, particularly where deliveries occur away from well-equipped facilities.

3.2. Reduce Neonatal Mortality

> *By 2030, end preventable deaths of newborns and children under 5 years of age, with all countries aiming to reduce neonatal mortality to at least as low as 12 per 1000 live births and under-5 mortality to at least as low as 25 per 1000 live births* (World Health Organization 2017).

The Global Burden of Disease (GBD) project found that 6.64 million deaths of children and adolescents occurred in 2017 (GBD 2019). In 2018, worldwide, 5.3 million, 85% of deaths among children, occurred in the first five years of life and 2.5 million (47%) in the first month of life (UNICEF 2019b). While this is a reduction on previous decades, it is still a huge public health burden. Infectious diseases remain the leading cause of preventable deaths for children under the age of 5 years. Lower respiratory tract infections (15%), diarrheal diseases (8%) and malaria (5%) are the leading causes of death globally among children under the age of 5 (UNICEF 2019b; Walker et al. 2013). Infants and children with undernutrition are particularly vulnerable to infections, and nutrition-related factors contribute to about 45% of deaths in children under 5 years of age (UNICEF 2019b).

In the original WHO monograph on nutrition and infection by Scrimshaw et al., the authors described the importance of breastfeeding on infant survival: "the fate of newborn infants in many pre-industrial areas seems to depend largely on whether they are breastfed or not - either they are nursed or they die" (Scrimshaw et al. 1968). The importance of exclusive breastfeeding for six months to reduce infant mortality was demonstrated in the classic WHO Collaborative Study, which demonstrated higher odds, as high as 6.0, of death before six months of age for infants who were not breastfed (WHO Collaborative Group 2000). In a review at the beginning of the millennium for UNICEF, exclusive breastfeeding during the first 6 months of life was found to have the single largest potential impact on child mortality of any preventive intervention (Jones et al. 2003). In this review, Level One evidence found that exclusive breastfeeding prevented infant deaths from diarrheal disease, lower respiratory tract disease and neonatal sepsis (Jones et al. 2003). In the PROBIT study, a postnatal breastfeeding intervention increased breastfeeding rates and reduced gastro-intestinal infections in infants (Kramer et al. 2001). Numerous studies and reviews have demonstrated the reduction in infectious diseases and hospital admissions of infants who are breastfed in countries at all levels of economic development (Horta and Victora 2013; Nguyen et al. 2020; U.S. Department of Health and Human Services 2011). The GBD study noted the importance of sudden infant death syndrome (SIDS), a condition for which exclusive breastfeeding offers some protection. A pooled data analysis of SIDS and breastfeeding found a reduction of 50% for infants breastfed for more than two months (Thompson et al. 2017).

Breastfeeding, according to the WHO recommendations, will result in a major reduction in the number of deaths under the age of 12 months by at least 820,000 each year (Rollins et al. 2016). It is also estimated that up to 50% of neonatal infant deaths could be averted if breastfeeding is initiated within the first hour of birth (Khan et al. 2015). In LMICs, an estimated 13% of all child deaths could be prevented if optimal breastfeeding levels are achieved (Nkoka et al. 2019).

The mechanism for the effectiveness of breastmilk in protecting against infections includes the composition of breastmilk with antibodies, non-specific immune factors and cellular components. An important mechanism for the prevention of infectious disease is the establishment of the microbiome. Breastmilk contains prebiotics and probiotics, and exclusive breastfeeding results in a 'healthy' microbiome. The microbiome of exclusively breastfed infants contains a higher proportion of Bifidobacteria and Lactobacillus spp., compared to formula-fed infants. These species are protective against gastrointestinal infections (Dong and Gupta 2019). The hormonal changes of breastfeeding contribute to the beneficial effects of breastfeeding on ovarian and breast cancers, and to birth spacing (Binns et al. 2016; Smith and Harvey 2011).

Additionally, breastfeeding reduces the risk of mortality in children under 5 years of age through its effect on birth spacing. Breastfeeding exerts a natural contraceptive effect by suppressing ovulation and delaying the return of menstruation, leading to increased birth spacing and fewer children (Sundhagen 2009). The risk of stunting, anemia, and mortality is higher among the children born to women with a higher number of births and short birth intervals.

Achievement of the SDG 3.2 target would require an annual reduction of approximately 1.5 million deaths of children under 12 months of age. At least 50% of this target could be realized by feeding infants according to the WHO Guidelines (Victora et al. 2016).

3.3. End Specified Epidemics

By 2030, end the epidemics of aids, tuberculosis, malaria and neglected tropical diseases and combat hepatitis, water-borne diseases and other communicable diseases (World Health Organization 2017).

Breastfeeding offers some protection against diarrheal diseases, reducing the rates by approximately 65% (U.S. Department of Health and Human Services 2011). It is not known if breastfeeding provides any protection against AIDS, malaria and tuberculosis, but partial protection is insufficient, and other public health measures are required. The WHO recommends that mothers known to be HIV-infected should be provided with lifelong antiretroviral therapy or antiretroviral prophylaxis interventions to reduce HIV transmission through breastfeeding. Mothers known to be HIV-infected (and whose infants are HIV-uninfected or of unknown HIV status) should exclusively breastfeed their infants for the first six months of life, introducing appropriate complementary foods thereafter, and then continuing to breastfeed (similar to the general population) (World Health Organization 2019b). Partial breastfeeding in the first six months of life provides inadequate protection against the vertical transmission of HIV from mother to child (Coutsoudis and Rollins 2003). Climate change will increase the global geographic spread and the incidence of many infectious diseases (Ahdoot et al. 2015; Philipsborn and Chan 2018). In infants and children, breastfeeding is important in protecting against infections, and making populations more resilient to such changes in the longer term (Lee and Binns 2019).

3.4. Reduce Premature Mortality from Non-Communicable Diseases

By 2030, reduce by one third premature mortality from non-communicable diseases through prevention and treatment, and promote mental health and well-being (World Health Organization 2017).

Improving breastfeeding rates has a role in the long-term prevention of non-communicable diseases. Within the time frame of the SDGs, breastfeeding can reduce the rates of some non-communicable diseases (NCDs) in mothers, and, generally, the effect is proportional to the total amount of time spent breastfeeding. Breastfeeding reduces the rates of ovarian cancer, premenopausal breast cancer and Type II diabetes (Gunderson et al. 2015; Scoccianti et al. 2015; Shield et al. 2018; Su et al. 2013; Victora et al. 2016; Zhang et al. 2004). A review of six cohort studies with 273,961 mothers found that the relative risk of developing diabetes for the highest duration of breastfeeding versus the lowest was 0.68 (95% CI = 0.57–0.82), reflecting a strong association between breastfeeding duration and lower rates of diabetes (Aune et al. 2014). Mothers who have breastfed may lose weight more rapidly after pregnancy and are less likely to be obese (Snyder et al. 2019). In a long-term follow up of overweight mothers, those who exclusively breastfed for four or more months and continued breastfeeding for 12 or more months were on average 8 kg lighter six years later compared to mothers who had not breastfed their infants (Sharma et al. 2014; Tahir et al. 2019). Breastfeeding reduces the risk of subsequent cardiovascular disease in mothers, by about 10% (Nguyen et al. 2017; Peters et al. 2017). The Women's Health Initiative Study (n = 139 681) found that a lifetime history of more than 12 months lactation resulted in a reduction of hypertension (OR = 0.88), diabetes (OR = 0.80), hyperlipidemia (Or = 0.81) and cardiovascular disease (OR = 0.91) (Schwarz et al. 2009). A follow-up of the European Investigation into Cancer and Nutrition prospective cohort (n = 322,972) found that mothers who had breastfed an infant had a reduced risk of dying over the following decade (OR = 0.92) (Merritt et al. 2015). The rate of perinatal depression is lower in mothers who continue breastfeeding (Xu et al. 2014; Yusuff et al. 2015).

The roots of non-communicable diseases lie in early life nutrition. The promotion of breastfeeding to meet the WHO guidance and appropriate nutrition during the first 1000 days of existence will result in a reduction in chronic disease later in life. Infants who are breastfed have lower rates of obesity. Other chronic diseases that are reduced by breastfeeding include diabetes (both type 1 and type 2), hypertension, cardiovascular disease, hyperlipidemia, some types of cancer and inflammatory bowel disease and related digestive system disorders (Ananthakrishnan et al. 2018; Binns et al. 2016).

The role of the human microbiome in health and disease is being increasingly recognized. Breastfeeding has an important role in establishing a healthy microbiome, and this contributes to the reduced prevalence of chronic disease through the developmental origins of health and disease process (Stiemsma and Michels 2018).

3.5. Strengthen the Prevention and Treatment of Substance Abuse

> *Strengthen the prevention and treatment of substance abuse, including narcotic drug abuse and harmful use of alcohol* (World Health Organization 2017).

The use of alcohol, tobacco products and other drugs is contra-indicated during breastfeeding, and this advice is given in public health guidelines for infant feeding (Binns et al. 2018; National Health and Medical Research Council 2013).

3.6. Halve Global Road Traffic Deaths and Injuries

By 2020, halve the number of global deaths and injuries from road traffic accidents (World Health Organization 2017).

Not applicable to breastfeeding.

3.7. Universal Access to Sexual and Reproductive Health-Care Services

By 2030, ensure universal access to sexual and reproductive health-care services, including for family planning, information and education, and the integration of reproductive health into national strategies and programmes (World Health Organization 2017).

Appropriate birth spacing results in healthier mothers and improved breastfeeding outcomes. Teenage mothers are less likely to breastfeed and to have a shorter duration of breastfeeding than older mothers (Kanhadilok et al. 2016). Postponing the age of reproduction until adulthood will improve the health of mothers and infants, and this highlights the need to make sexual health services available to this age group.

Temporary contraception for new mothers whose monthly bleeding has not returned requires exclusive or full breastfeeding day and night of an infant less than 6 months old and may be 98–99% effective (Tucker et al. 2011; World Health Organization 2018a).

3.8. Universal Health Coverage

Achieve universal health coverage, including financial risk protection, access to quality essential health-care services and access to safe, effective, quality and affordable essential medicines and vaccines for all (World Health Organization 2017).

Universal health care provides the opportunity for mothers and fathers to attend health services and obtain education needed before birth. After birth mothers need appropriate treatment and advice on the best ways to continue breastfeeding if complications occur, such as mastitis. On the 30th anniversary of the Convention of the Rights of the Child the UNICEF Special Rapporteur on the Right to Food issued a report stating that support for breastfeeding, as implied in Article 18, is consistent with the legal obligations most countries have committed to, by their ratification of the Convention, which includes obligations to protect children's right to a healthy food environment (United Nations Children's Fund and United Nations Special Rapporteur on the Right to Food 2018).

Breastfeeding will reduce the demand on health services as infants; children and adults are less likely to require health service treatment (see above). For mothers, breastfeeding reduces the rates of a significant number of diseases, reducing overall demand and enabling expansion to unreached areas and releasing resources for promoting and supporting breastfeeding.

3.9. Reduce Death and Illness from Hazardous Chemicals, Pollution and Contamination

> *By 2030, substantially reduce the number of deaths and illnesses from hazardous chemicals and air, water and soil pollution and contamination* (World Health Organization 2017).

Persistent organic pollutants (POPs) have been of environmental and health concern for more than half a century. The only two countries with long temporal trend studies are Japan and Sweden. In most cases, the trends show decreasing concentrations of POPs in mothers' milk (Fang et al. 2015). Plastic infant feeding bottles have been shown to release bisphenol A and S into infant formula (Russo et al. 2018; Walker et al. 2011). Obviously, breastfeeding is the simplest way to avoid this problem.

4. Breastfeeding and the Other SDGs

The Sustainable Development Goals are an integrated whole, and Goal 3 should be considered along with the other 16 goals. Increasing breastfeeding is essential to achieving Goal 3 of the SDGs. Breastfeeding will also make a major contribution to achieving other goals:

- SDG 1: No Poverty. There is a strong link between health and economic development. Gallardo-Albarran has examined progress in the 20th century and demonstrated that improving health has contributed to a narrowing of the income gap between countries (Gallardo-Albarran 2018). In a large long-term cohort study, breastfeeding was associated with improved performance in intelligence tests three decades later, and there was an association between breastfeeding and increased educational attainment and income in adulthood (Victora et al. 2015).
- SDG 2: Zero Hunger. Breastmilk, while providing the optimal source of nutrition for infants, and substantial benefits for mother, is also substantially cheaper than the available alternative, which is commercial infant formula. Breastmilk represents a valuable lost 'natural' resource when replaced with formula. It was estimated that, in 2012, the total potential value of breastmilk in the USA was 110

billion USD, but 2/3 of this is lost due to sub-optimal breastfeeding (Smith 2013). The total world milk formula sales grew by 40.8% from 5.5 to 7.8kg/infant/child in the period 2008–2013, and is projected to grow at the rate of 9% per annum over the next decade (Allied Market Research 2017; Baker et al. 2016).

- SDG 4: Education. There is increasing evidence that breastfeeding also confers long-term benefits to the infant, not only in health and disease, but also in the development of human capabilities (Horta 2019; Strom et al. 2019). Children who have been breastfed will have, on average, a slightly higher IQ and be healthier, including being less likely to be obese. These factors will improve education performance.

- SDG 5: End Discrimination against Women. Promoting breastfeeding should prevent 20,000 deaths of mothers from breast cancer annually (Aryeetey et al. 2018; Victora et al. 2016). As part of the strategies to enable breastfeeding, the provision of paid parental leave should be expanded (Heymann et al. 2017). Mothers are often unable to continue breastfeeding their infants after their return to work, but there are many strategies to make this easier for them (Lakati et al. 2002).

- SDG 8: Decent Work and Economic Growth. Breastfed infants had improved cognitive development (higher IQ), and this appears to be dose-related. They are also healthier, and the advantage of better health persists into adulthood. When they reach adulthood, a large cohort study documented higher incomes in breastfed infants (Straub et al. 2016; Victora et al. 2015). In situations when infants are fed formula, the cost and time spent in preparation impose significant burdens on a family (Siregar et al. 2018; Smith 2019; Walters et al. 2019).

- SDG 9: Reduced Inequality. The breastfeeding of all infants will assist in reducing inequality. Where artificial infant formula is promoted and available, poorer families may purchase cheaper varieties, over dilute the formula to reduce costs or use contaminated water supplies. In recent years, there have been instances of contamination of infant formula and shortages of supplies (Qiu et al. 2010; Xin and Stone 2008). It is difficult (but not impossible) for mothers to resume breastfeeding, and in the case of the Chinese contamination, parents spent large amounts of money buying formula that had been imported from high income countries. Food security for infants and young children is not possible without high rates of breastfeeding (Salmon 2015).

- SDG 12: Responsible Consumption and Production. The alternative to breastmilk is the use of infant formula, which is usually based on dairy milk production. The carbon footprint of milk production is high and manufacture into infant formula requires large quantities of potable water in the order of 4000 L of water for one kg of formula milk powder (Clune et al. 2017; Hagemann et al. 2011; Karlsson et al. 2019; Rollins et al. 2016). If access to pure water is difficult, and it may become more difficult with climate change, using formula can be very dangerous. The cost of less than optimal breastfeeding is substantial, due to the increase in child and maternal morbidity and mortality (Bartick and Reinhold 2010; Bartick et al. 2013). The loss of productivity due to the lower level of cognitive development in infants who are not breastfed is estimated at 302 billion USD annually, or 0.49% of world gross national income (Rollins et al. 2016).

5. Disaster Management and Breastfeeding

The 17 Sustainable Development Goals (SDGs) include 169 targets used to monitor achievements. The United Nations Office for Disaster Risk Reduction (UNISDR) has identified 25 targets related to disaster risk reduction in 10 of the 17 SDGs (United Nations Office for Disaster Risk Reduction 2015). Unfortunately, with the effects of climate change becoming more prominent, the number and severity of natural disasters is increasing. It is important that, in any disaster situation requiring the provision of food, the breastfeeding of infants and young children receives early attention. If it is at all possible for mothers to continue to breastfeed, they should be encouraged as much as possible, in order to minimize infant morbidity. The provision of emergency supplies of infant formula in disaster situations should be carefully managed to avoid permanent changes to infant feeding patterns. Only high-quality infant formula in plain packaging should be used to avoid advertising and breaching the International Code of Marketing of Breast-Milk Substitutes (the 'BMS Code') (WHO 1981; World Health Organization 2018b). The promotion of breastfeeding and feeding mothers (water and food) will be more sustainable and provide better health outcomes than distributing infant formula (Binns et al. 2012b; Summers and Bilukha 2018).

6. Achieving Breastfeeding Goals

In historical times, breastfeeding was universal, as all infants were breastfed by their mothers or a wet nurse. Currently, the WHO/UNICEF estimates that worldwide, 44% of infants begin breastfeeding within one hour of birth (UNICEF 2019a). Exclusive breastfeeding 0–6 months is estimated at 42%, and 65% continue beyond 12 months, but these results are dependent on the method and definitions used in collecting the data (Binns et al. 2012a). Improving breastfeeding rates is one of the most important steps in achieving the targets of SDG 3 (Perez-Escamilla

2017). The Sustainable Development Goals provide an appropriate framework for supporting universal breastfeeding targets. The WHO and UNICEF have endorsed many strategies to improve breastfeeding that have been evaluated. However, the resources and the will to implement have not been forthcoming. In part, this is due to the long-term nature of public health benefits compared to the short-term political cycle, and, in part, due to the role of industry in infant formula production and promotion. Strategies to achieve optimum breastfeeding will require the involvement of family, health services, local and national governments and society in general. A global environment supportive of breastfeeding is required, which would include the following components.

6.1. A Supportive Societal Environment

Media reports, advertising and social media ideally should all portray the benefits and joys of breastfeeding, as has already being done by many mothers' groups and breastfeeding advocates. Breastfeeding needs to be regarded as the norm and specific programs of information need to be implemented to provide accurate information. While breastfeeding is an individual activity, it requires investment at a societal level (Brown 2017). In a globally connected world, countries which are large producers of infant formula have a duty to limit their exports of these products (Galtry 2013; Gribble and Smith 2014). Public places need facilities where mothers can care for their infants and feel comfortable about breastfeeding. In many economies, mothers undertake paid employment and would like to continue breastfeeding while returning to work. Maternal leave, including paid leave, has been shown to improve breastfeeding rates (Chai et al. 2018). Informal arrangements to breastfeed and return to employment are also beneficial (Lakati et al. 2002).

6.2. Maternal and Family Preparation for Breastfeeding

Giving mothers knowledge and motivation about breastfeeding is important. The best time to begin to discuss the benefits of breastfeeding is before or soon after the mother becomes pregnant (National Health and Medical Research Council 2012). Antenatal education, counselling and encouragement are important, and are of benefit. This can be achieved through formal structures, the use of volunteer health promotors or village health workers. Fathers and other close family members are important participants in infant care, and they all need knowledge and motivation (Maycock et al. 2013). The use of smartphones has become almost universal and they provide an important medium for sharing knowledge and experience, provided that this is moderated to ensure accuracy (White et al. 2019).

6.3. Health Service Support

Universal health access will mean that all health workers will need education into the benefits of breastmilk and the BMS Marketing Code so that BFHI principles can be implemented at all levels of the health service. Breastfeeding counselling and promotion during antenatal education is important. A Cochrane review on prenatal intervention found that healthcare-professional-led breastfeeding education, and non-healthcare-professional-led counselling and peer support interventions can result in some improvements in the number of women beginning to breastfeed (Balogun et al. 2016; Perez-Escamilla 2020). The Baby Friendly Hospital (and health service) Initiative needs to be implemented universally to provide a favorable environment for mothers and infants (Perez-Escamilla 2020; Spaeth et al. 2018; WHO 2019; WHO UNICEF 2018). At present, the endorsement or even accreditation of BFHI is widespread, but may not always be applied in practice (Hawkins et al. 2014; WHO UNICEF 2018). The aims of the BFHI include the early initiation of breastfeeding (early skin to skin contact), feeding colostrum and continued exclusive breastfeeding, all of which are highly effective and low-cost interventions for short and long term health of infants (Bhutta and Labbok 2011; Bhutta et al. 2014; Rollins and Doherty 2019; Rollins et al. 2016). Hospitals and other components of the health system must not promote infant formula in any way, or accept free or low-cost supplies or promotional material.

6.4. The Promotion of Breast Milk Substitutes

Recognizing the damage caused by breast milk substitutes, the World Health Assembly approved the International Code of Marketing of Breast-Milk Substitutes in 1981(WHO 2020a). Support, in principle, for the code is extensive, but, when it comes to legislative action, it is often limited (WHO and UNICEF 2020). Very few countries have marketing restrictions on products suitable for up to three years of age, and the majority have voluntary restrictions only up to six months of age (WHO 2020a, 2020b). The Code is widely breached in many LMICs, with widespread advertising across all media, including television, print and social media (IBFAN 2018). Infant formula products designed for toddlers are widely marketed, and parents often assume that the exaggerations in the advertisements apply to infants of all ages. Many countries, particularly middle-income countries, have experienced a surge in infant formula use, as incomes increase (Neves et al. 2020; Son 2017). Vietnam is an example of an LMIC that has experienced sales growth of 12% per annum over the past decade (Whitehead 2020). The government's response has been similar to other Asian countries such as China and Japan, to encourage and support local production of infant formula. This is despite cow milk production for infant formula having a large carbon footprint and requiring large quantities of potable water in its production (Smith 2019; Sultana et al. 2014). The use of plain label packaging has been highly successful in reducing tobacco sales, and could be beneficial in curbing the widespread promotion of infant formula (Drovandi et al. 2019; Wise 2017).

6.5. Specific Groups May Need Special Programs eg Obese Mothers, Illness

There are a small number of mothers for whom breastfeeding may be contra-indicated (World Health Organization 2009). These mothers and other mother–infant dyads in which there are physical conditions (e.g., obesity), or illnesses that impair breastfeeding, may require special counselling and support from the health system. Breastfeeding can usually be maintained using the mother's own supply or from donor breastmilk.

6.6. The COVID-19 Pandemic

There have been reports that mothers with COVID have stopped breastfeeding their infants, and this action is being recommended in some countries, including Japan. However, the World Health Organisation recommends: "that mothers with suspected or confirmed COVID-19 should be encouraged to initiate or continue to breastfeed (World Health Organization 2020). Mothers should be counselled that the benefits of breastfeeding substantially outweigh the potential risks for transmission". This is the position of most pediatric societies (Rozycki and Kotecha 2020).

7. Conclusions

Globally, appropriate breastfeeding practices have the potential to prevent over 800,000 deaths of children under five years of age annually. In low income countries, an estimated 13% of all child deaths could be prevented if optimal levels of breastfeeding were achieved. If breastfeeding increases, infants, children and adults will all be healthier, and contribute to achieving the SDGs. Minimizing avoidable health care costs will facilitate achieving universal health care. Promoting breastfeeding will not have any detrimental on any of the other SDGs.

The sustainability costs of breastmilk substitutes promoted by industry as alternatives to breastfeeding are massive, and from an economic perspective, the cost of not breastfeeding is into the trillions of dollars. When we add the human costs of not breastfeeding, the benefits of promoting breastfeeding, as the world works towards achieving the Sustainable Development Goals, are overwhelming. To achieve these benefits, the combined efforts of many sectors of society and government will be required.

We summarize our chapter in the following statement:

Breastmilk is best for all babies, mother–baby dyads, for lifetime health, for sustainability, and for the environment and economy. Protect it from harm or it is gone.

Author Contributions: Each author made substantial contributions to the conception of the chapter, the acquisition, analysis, or interpretation of the literature references used in the work, and writing and revision. The final manuscript has been approved by all authors, and we agree to be personally accountable for the authors' contributions

Funding: This research received no external funding.

Conflicts of Interest: The authors declare no conflict of interest.

References

Ahdoot, Samantha, Susan. E. Pacheco, and The Council on Environmental Health American Academy of Pediatrics. 2015. Global Climate Change and Children's Health. *Pediatrics* 136: E1468–E1484. [CrossRef] [PubMed]

Allied Market Research. 2017. Baby Infant Formula Market Overview. Available online: https://www.alliedmarketresearch.com/baby-infant-formula-market (accessed on 5 October 2019).

Ananthakrishnan, Ashwin N., Charles. N. Bernstein, Dimitros Iliopoulos, Andrew Macpherson, Markus. F. Neurath, Raja. A. R. Ali, Stephan R. Vavricka, and Claudio Fiocchi. 2018. Environmental triggers in IBD: A review of progress and evidence. *Nature Reviews Gastroenterology & Hepatology* 15: 39–49. [CrossRef]

Aryeetey, Richmond, Amber Hromi-Fiedler, Seth Adu-Afarwuah, Esi Amoaful, Gifty Ampah, Marian Gatiba, Akosua Kwakye, Gloria Otoo, Gyikua Plange-Rhule, Isabella Sagoe-Moses;, and et al. 2018. Pilot testing of the Becoming Breastfeeding Friendly toolbox in Ghana. *International Breastfeeding Journal* 13: 30. [CrossRef] [PubMed]

Aune, Dagfinn, Teresa Norat, Pal Romundstad, and Lars. J. Vatten. 2014. Breastfeeding and the maternal risk of type 2 diabetes: a systematic review and dose-response meta-analysis of cohort studies. *Nutrition, Metabolism and Cardiovascular Diseases* 24: 107–15. [CrossRef] [PubMed]

Baker, Phillip, Julie Smith, Libby Salmon, Sharon Friel, George Kent, Alessandro Iellamo, J P Dadhich, and Mary J Renfrew. 2016. Global trends and patterns of commercial milk-based formula sales: is an unprecedented infant and young child feeding transition underway? *Public Health Nutrition* 19: 2540–50. [CrossRef]

Balogun, Olukunmi O, Elizabeth J OSullivan ', Alison McFadden, Erika Ota, Anna Gavine, Christine D Garner, Mary J Renfrew, and Stephen MacGillivray. 2016. Interventions for promoting the initiation of breastfeeding. *Cochrane Database of Systematic Reviews* 11: CD001688. [CrossRef] [PubMed]

Bartick, Melissa, and Arnold Reinhold. 2010. The Burden of Suboptimal Breastfeeding in the United States: A Pediatric Cost Analysis. *Pediatrics* 125: E1048–E1056. [CrossRef]

Bartick, Melissa C., Alison M. Stuebe, Eleanor Bimla Schwarz, Christine Luongo, Arnold G. Reinhold, and E. Michael Foster. 2013. Cost Analysis of Maternal Disease Associated With Suboptimal Breastfeeding. *Obstetrics and Gynecology* 122: 111–19. [CrossRef]

Bhutta, Zulfiqar A, and Miriam Labbok. 2011. Scaling up breastfeeding in developing countries. *Lancet* 378: 378–80. [CrossRef]

Bhutta, Zulfiqar A., Jai K. Das, Rajiv Bahl, Joy E. Lawn, Rehana A. Salam, Vinod K. Paul, M. Jeeva Sankar, Hannah Blencowe, Arjumand Rizvi, Victoria B. Chou;, and et al. 2014. Can available interventions end preventable deaths in mothers, newborn babies, and stillbirths, and at what cost? *The Lancet* 384: 347–70. [CrossRef]

Binns, Colin W. 1976. Food, sickness and death in children of the highlands of Papua, New Guinea. *Journal of Tropical Pediatrics* 22: 9–11. [CrossRef] [PubMed]

Binns, Colin, Andy Lee, Kay Sauer, and Katie Hewitt. 2012a. Reported Breastfeeding Rates in the Asia-Pacific Region. *Current Pediatric Reviews* 8: 339–45. [CrossRef]

Binns, Colin W, Mi Kyung Lee, Li Tang, Chuan Yu, Tomiko Hokama, and Andy Lee. 2012b. Ethical issues in infant feeding after disasters. *Asia Pacific Journal of Public Health* 24: 672–80. [CrossRef] [PubMed]

Binns, Colin W., Mi Kyung Lee, and Wah Yun Low. 2016. The Long-Term Public Health Benefits of Breastfeeding. *Asia Pacific Journal of Public Health* 28: 7–14. [CrossRef]

Binns, Colin, Mi Kyung Lee, and Masaharu Kagawa. 2017. Ethical Challenges in Infant Feeding Research. *Nutrients*, 9. [CrossRef]

Binns, Colin, Mi Kyung Lee, Masaharu Kagawa, Wah Yun Low, Jane Scott, Andy Lee, Alfred Zerfas, Bruce Maycock, Liqian Qiu, Aza Yusuff;, and et al. 2018. Infant Feeding Guidelines for the Asia Pacific Region. *Asia Pacific Journal of Public Health* 40: 1010539518809823. [CrossRef]

Brown, Amy. 2017. Breastfeeding as a public health responsibility: A review of the evidence. *Journal of Human Nutrition and Dietetics* 30: 759–70. [CrossRef]

Chai, Yan, Arijit Nandi, and Jody Heymann. 2018. Does extending the duration of legislated paid maternity leave improve breastfeeding practices? Evidence from 38 low-income and middle-income countries. *BMJ Global Health*, 3. [CrossRef]

Clune, Stephen, Enda Crossin, and Karli Verghese. 2017. Systematic review of greenhouse gas emissions for different fresh food categories. *Journal of Cleaner Production* 140: 766–83. [CrossRef]

Coutsoudis, Anna, and Nigel Rollins. 2003. Breast-feeding and HIV transmission: The jury is still out. *Journal of Pediatric Gastroenterology and Nutrition* 36: 434–42. [CrossRef]

Dong, Tien S., and Arpana Gupta. 2019. Influence of Early Life, Diet, and the Environment on the Microbiome. *Clinical Gastroenterology and Hepatology* 17: 231–42. [CrossRef] [PubMed]

Drovandi, Aaron, Peta-Ann Teague, Beverly Glass, and Bunmi Malau-Aduli. 2019. A systematic review of the perceptions of adolescents on graphic health warnings and plain packaging of cigarettes. *Systematic Reviews* 8: 25. [CrossRef] [PubMed]

Fang, Johan, Elisabeth Nyberg, Ulrika Winnberg, Anders Bignert, and Ake Bergman. 2015. Spatial and temporal trends of the Stockholm Convention POPs in mothers' milk - a global review. *Environmental Science and Pollution Research* 22: 8989–9041. [CrossRef]

Gallardo-Albarran, Daniel. 2018. Health and economic development since 1900. *Economics & Human Biology* 31: 228–37. [CrossRef]

Galtry, Judith. A. 2013. Improving the New Zealand dairy industry's contribution to local and global wellbeing: the case of infant formula exports. *New Zealand Medical Journal* 126: 82–89. [PubMed]

Global Burden of Diseases (GBD) 2017 Child and Adolescent Health Collaborators. 2019. Diseases, Injuries, and Risk Factors in Child and Adolescent Health, 1990 to 2017: Findings From the Global Burden of Diseases, Injuries, and Risk Factors 2017 Study. *JAMA Pediatr* 173: e190337. [CrossRef]

Gribble, Karleen, and Julie. P. Smith. 2014. China's 'White Gold' Infant Formula Rush Comes at a Public Health Cost. Available online: http://theconversation.com/chinas-white-gold-infant-formula-rush-comes-at-a-public-health-cost-34363 (accessed on 13 Sept 2020).

Gunderson, Erica P., Shanta R. Hurston, Xian Ning, Joan C. Lo, Yvonne Crites, David Walton, Kathryn G. Dewey, Robert A. Azevedo, Stephen Young, Gary Fox;, and et al. 2015. Lactation and Progression to Type 2 Diabetes Mellitus After Gestational Diabetes Mellitus. *Annals of Internal Medicine* 163: 889. [CrossRef] [PubMed]

Hagemann, Martin, Torsten Hemme, Asaah Ndambi, Othman Alqaisi, and Mst Nadira Sultana. 2011. Benchmarking of greenhouse gas emissions of bovine milk production systems for 38 countries. *Animal Feed Science and Technology* 166–67: 46–58. [CrossRef]

Hawkins, Summer Sherburne, Ariel Dora Stern, Christopher F. Baum, and Matthew W. Gillman. 2014. Compliance with the Baby-Friendly Hospital Initiative and impact on breastfeeding rates. *Archives of Disease in Childhood-Fetal and Neonatal Edition* 99: F138–F43. [CrossRef]

Heymann, Jody, Aleta R. Sprague, Arijit Nandi, Alison Earle, Priya Batra, Adam Schickedanz, Paul J. Chung, and Amy Raub. 2017. Paid parental leave and family wellbeing in the sustainable development era. *Public Health Reviews* 38: 21. [CrossRef]

Horta, Bernado L. 2019. Breastfeeding: Investing in the Future. *Breastfeeding Medicine* 14: S11–S12. [CrossRef] [PubMed]

Horta, Bernado L., and Cesar G. Victora. 2013. The Short Term Effects of Breastfeeding: A Systematic Review. Available online: http://www.who.int/maternal_child_adolescent/documents/breastfeeding_short_term_effects/en/ (accessed on 1 February 2018).

IBFAN. 2018. *Report on the Monitoring of the Code in 11 Countries of Asia*. Georgetown: International Baby Food Action Network, Available online: https://www.bpni.org/wp-content/uploads/2018/12/Monitoring-of-the-Code-in-11-Countries-of-Asia.pdf (accessed on 10 July 2020).

Jones, Gareth, Richard W. Steketee, Robert E. Black, Zulfiqar A. Bhutta, Saul S. Morris, and Bellagio Child Survival Study Group. 2003. How many child deaths can we prevent this year? *Lancet* 362: 65–71. [CrossRef]

Kanhadilok, Supannee, Nancy L. McCain, Jacqueline M. McGrath, Nancy Jallo, Sarah K. Price, and Chantira Chiaranai. 2016. Factors Associated With Exclusive Breastfeeding Through Four Weeks Postpartum in Thai Adolescent Mothers. *The Journal of Perinatal Education* 25: 150–61. [CrossRef] [PubMed]

Karlsson, Johan O., Tara Garnett, Nigel C. Rollins, and Elin Roos. 2019. The carbon footprint of breastmilk substitutes in comparison with breastfeeding. *Journal of Cleaner Production* 222: 436–45. [CrossRef] [PubMed]

Khan, Jehangir, Linda Vesel, Rajiv Bahl, and José Carlos Martines. 2015. Timing of Breastfeeding Initiation and Exclusivity of Breastfeeding During the First Month of Life: Effects on Neonatal Mortality and Morbidity-A Systematic Review and Meta-analysis. *Maternal and Child Health Journal* 19: 468–79. [CrossRef]

Kramer, Michael S., Beverley Chalmers, Ellen D. Hodnett, Zinaida Sevkovskaya, Irina Dzikovich, Stanley Shapiro, Jean-Paul Collet, Irina Vanilovich, Irina Mezen, Thierry Ducruet, and et al. 2001. Promotion of Breastfeeding Intervention Trial (PROBIT): A randomized trial in the Republic of Belarus. *JAMA* 285: 413–20. [CrossRef]

Lakati, Alice, Colin Binns, and Mark Stevenson. 2002. Breast-feeding and the working mother in Nairobi. *Public Health Nutrition* 5: 715–18. [CrossRef]

Lee, MiKyung, and Colin W. Binns. 2019. Breastfeeding and the Risk of Infant Illness in Asia: A Review. *International Journal of Environmental Research and Public Health* 17: 186. [CrossRef]

Maastrup, Ragnhild, Laura N. Haiek, and Neo-BFHI Survey Group. 2019. Compliance with the "Baby-friendly Hospital Initiative for Neonatal Wards" in 36 countries. *Maternal & Child Nutrition* 15: e12690. [CrossRef]

Maycock, Bruce, Colin W. Binns, Satvinder Dhaliwal, Jenny Tohotoa, Yvonne Hauck, Sharyn Burns, and Peter Howat. 2013. Education and support for fathers improves breastfeeding rates: A randomized controlled trial. *Journal of Human Lactation* 29: 484–90. [CrossRef]

Merritt, Melissa A., Elio Riboli, Neil Murphy, Mai Kadi, Anne Tjonneland, Anja Olsen, Kim Overvad, Laure Dossus, Laureen Dartois, Françoise Clavel-Chapelon;, and et al. 2015. Reproductive factors and risk of mortality in the European Prospective Investigation into Cancer and Nutrition; a cohort study. *BMC Medicine* 13: 252. [CrossRef] [PubMed]

National Health and Medical Research Council. 2012. *Infant Feeding Guidelines.* Canberra: National Health and Medical Research Council.

National Health and Medical Research Council. 2013. *Infant Feeding Guidelines for Health Workers.* Canberra: NHMRC.

Neves, Paulo A. R., Giovanna Gatica-Dominguez, Nigel C. Rollins, Ellen Piwoz, Phillip Baker, Aluísio J. D. Barros, and Cesar G. Victora. 2020. Infant Formula Consumption Is Positively Correlated with Wealth, Within and Between Countries: A Multi-Country Study. *Journal of Nutrition* 150: 910–17. [CrossRef]

Nguyen, Binh, Kai Jin, and Ding Ding. 2017. Breastfeeding and maternal cardiovascular risk factors and outcomes: A systematic review. *PLoS ONE* 12: e0187923. [CrossRef] [PubMed]

Nguyen, Phung, Colin W. Binns, Anh Vo Van Ha, Tan Khac Chu, Luat Cong Nguyen, Dat Van Duong, Dung Van Do, and Andy H. Lee. 2020. Prelacteal and early formula feeding increase risk of infant hospitalisation: a prospective cohort study. *Archives of Disease in Childhood* 105: 122–26. [CrossRef]

Nkoka, Own, Peter A. M. Ntenda, Victor Kanje, Edith B. Milanzi, and Amit Arora. 2019. Determinants of timely initiation of breast milk and exclusive breastfeeding in Malawi: A population-based cross-sectional study. *International Breastfeeding Journal,* 14. [CrossRef]

Perez-Escamilla, Rafael. 2017. Food Security and the 2015–2030 Sustainable Development Goals: From Human to Planetary Health: Perspectives and Opinions. *Current Developments in Nutrition* 1: e000513. [CrossRef] [PubMed]

Perez-Escamilla, Rafael. 2020. Breastfeeding in the 21st century: How we can make it work. *Social Science & Medicine* 244: 112331. [CrossRef]

Peters, Sanne A. E., Ling Yang, Yu Guo, Yiping P. Chen, Zheng Bian, Jianwei Du, Jie Yang, Shanpeng Li, Liming Li, Mark Woodward, and et al. 2017. Breastfeeding and the Risk of Maternal Cardiovascular Disease: A Prospective Study of 300 000 Chinese Women. *Journal of the American Heart Association,* 6. [CrossRef]

Philipsborn, Rebecca P., and Kevin Chan. 2018. Climate Change and Global Child Health. *Pediatrics,* 141. [CrossRef]

Qiu, Liqian, Colin W. Binns, Yun Zhao, Andy H. Lee, and Xing Xie. 2010. Breastfeeding practice in Zhejiang province, PR China, in the context of melamine-contaminated formula milk. *Journal of Health, Population, and Nutrition* 28: 189–98.

Raheem, Raheema Abdul, Colin W. Binns, Hui Jun Chih, and Kay Sauer. 2014. Determinants of the introduction of prelacteal feeds in the Maldives. *Breastfeeding Medicine* 9: 473–78. [CrossRef] [PubMed]

Rollins, Nigel, and Tanya Doherty. 2019. Improving breastfeeding practices at scale. *Lancet Global Health* 7: E292–E293. [CrossRef]

Rollins, Nigel C., Nita Bhandari, Nemat Hajeebhoy, Susan Horton, Chessa K. Lutter, Jose C. Martines, Ellen G. Piwoz, Linda M. Richter, Cesar G. Victora, and Group Lancet Breastfeeding Series. 2016. Why invest, and what it will take to improve breastfeeding practices? *Lancet* 387: 491–504. [CrossRef]

Rozycki, Henry J., and Sailesh Kotecha. 2020. Covid-19 in pregnant women and babies: What pediatricians need to know. *Paediatric Respiratory Reviews*. [CrossRef]

Russo, Giacomo, Francesco Barbato, Eleonora Cardone, Margherita Fattore, Stefania Albrizio, and Lucia Grumetto. 2018. Bisphenol A and Bisphenol S release in milk under household conditions from baby bottles marketed in Italy. *Journal of Environmental Science and Health Part B-Pesticides Food Contaminants and Agricultural Wastes* 53: 116–20. [CrossRef]

Salmon, Libby. 2015. Food security for infants and young children: An opportunity for breastfeeding policy? *International Breastfeeding Journal*. [CrossRef]

Schwarz, Eleanor Bimla, Roberta M. Ray, Alison M. Stuebe, Matthew A. Allison, Roberta B. Ness, Matthew S. Freiberg, and Jane A. Cauley. 2009. Duration of lactation and risk factors for maternal cardiovascular disease. *Obstet Gynecol* 113: 974–82. [CrossRef]

Scoccianti, Chiara, Timothy J. Key, Annie S. Anderson, Paola Armaroli, Franco Berrino, Michele Cecchini, Marie-Christine Boutron-Ruault, Mchael Leitzmann, Teresa Norat, Hilary Powers;, and et al. 2015. European Code against Cancer 4th Edition: Breastfeeding and cancer. *Cancer Epidemiology* 39: S101–S106. [CrossRef]

Scrimshaw, Nevin Stewart, Carl Ernest Taylor, and John Everett Gordon. 1968. Interactions of nutrition and infection. *Monograph Series World Health Organ* 57: 3–329.

Sentilhes, Loïc, Benjamin Merlot, Hugo Madar, François Sztark, Stéphanie Brun, and Catherine Deneux-Tharaux. 2016. Postpartum haemorrhage: prevention and treatment. *Expert Review of Hematology* 9: 1043–61. [CrossRef]

Sharma, Andrea J., Doborah L. Dee, and Samantha M. Harden. 2014. Adherence to breastfeeding guidelines and maternal weight 6 years after delivery. *Pediatrics* 134: S42–S49. [CrossRef]

Shield, Kevin D., Laure Dossus, Agnès Fournier, Claire M. Micallef, Sabina Rinaldi, Agnès Rogel, Isabelle Heard, Sophie Pilleron, Freddie Bray, and Isabelle Soerjomataram. 2018. The impact of historical breastfeeding practices on the incidence of cancer in France in 2015. *Cancer Causes & Control* 29: 325–32. [CrossRef]

Siregar, Adiatma Y. M., Pipit Pitriyan, and Dylan Walters. 2018. The annual cost of not breastfeeding in Indonesia: the economic burden of treating diarrhea and respiratory disease among children (<24mo) due to not breastfeeding according to recommendation. *International Breastfeeding Journal*, 13. [CrossRef]

Smith, Julie P. 2013. "Lost Milk?" Counting the Economic Value of Breast Milk in Gross Domestic Product. *Journal of Human Lactation* 29: 537–46. [CrossRef] [PubMed]

Smith, Julie P. 2019. A commentary on the carbon footprint of milk formula: harms to planetary health and policy implications. *International Breastfeeding Journal* 14: 49. [CrossRef]

Smith, Julie. P., and Peta. J. Harvey. 2011. Chronic disease and infant nutrition: Is it significant to public health? *Public Health Nutrition* 14: 279–89. [CrossRef]

Snyder, Gabrielle G., Claudia Holzman, Tao Sun, Bertha L. Bullen, Marianne Bertolet, and Janet M. Catov. 2019. Breastfeeding Greater Than 6 Months Is Associated with Smaller Maternal Waist Circumference Up To One Decade After Delivery. *Journal of Womens Health* 28: 462–72. [CrossRef]

Son, Minh. 2017. Vietnam's appetite for foreign baby formula is making US giants rich. *VN Express*, May 26.

Spaeth, Anna, Elisabeth Zemp, Sonja Merten, and Julia Dratva. 2018. Baby-Friendly Hospital designation has a sustained impact on continued breastfeeding. *Maternal and Child Nutrition*, 14. [CrossRef]

Stiemsma, Leah T., and Karin B. Michels. 2018. The Role of the Microbiome in the Developmental Origins of Health and Disease. *Pediatrics*, 141. [CrossRef] [PubMed]

Straub, Niels, Philipp Grunert, Kate Northstone, and Pauline Emmett. 2016. Economic impact of breast-feeding-associated improvements of childhood cognitive development, based on data from the ALSPAC. *British Journal of Nutrition*, 1–6. [CrossRef]

Strom, Marin, Erik Lykke Mortensen, Ulrik Schiøler Kesmodel, Thorhallur Halldorsson, Jørn Olsen, and Sjurdur F. Olsen. 2019. Is breast feeding associated with offspring IQ at age 5? Findings from prospective cohort: Lifestyle During Pregnancy Study. *BMJ Open*, 9. [CrossRef] [PubMed]

Su, Dada, Maria Pasalich, Andy H. Lee, and Colin W. Binns. 2013. Ovarian cancer risk is reduced by prolonged lactation: A case-control study in southern China. *The American Journal of Clinical Nutrition* 97: 354–59. [CrossRef] [PubMed]

Sultana, Madira, Mohammad Mohi Uddin, Bradley G. Ridoutt, and Kurt J. Peters. 2014. Comparison of water use in global milk production for different typical farms. *Agricultural Systems* 129: 9–21. [CrossRef]

Summers, Aimee, and Oleg O. Bilukha. 2018. Suboptimal infant and young child feeding practices among internally displaced persons during conflict in eastern Ukraine. *Public Health Nutrition* 21: 917–26. [CrossRef]

Sundhagen, Rebecca. 2009. Breastfeeding and Child Spacing. In *Childbirth Across Cultures: Ideas and Practices of Pregnancy, Childbirth and the Postpartum*. Edited by H. Selin. Dordrecht: Springer Netherlands, pp. 23–32.

Tahir, Muna J., Jacob L. Haapala, Laurie P. Foster, Katy M. Duncan, April M. Teague, Elyse O. Kharbanda, Patricia M. McGovern, Kara M. Whitaker, Kathleen M. Rasmussen, David A. Fields, and et al. 2019. Association of Full Breastfeeding Duration with Postpartum Weight Retention in a Cohort of Predominantly Breastfeeding Women. *Nutrients* 11: 938. [CrossRef]

Thompson, John M. D., Kawai Tanabe, Rachel Y. Moon, Edwin A. Mitchell, Cliona McGarvey, David Tappin, Peter S. Blair, and Fern R. Hauck. 2017. Duration of Breastfeeding and Risk of SIDS: An Individual Participant Data Meta-analysis. *Pediatrics*, 140. [CrossRef] [PubMed]

Tucker, Christine M., Ellen K. Wilson, and Ghazaleh Samandari. 2011. Infant feeding experiences among teen mothers in North Carolina: Findings from a mixed-methods study. *International Breastfeeding Journal* 6: 14. [CrossRef]

U.S. Department of Health and Human Services. 2011. *The Surgeon General's Call to Action to Support Breastfeeding*. Washington, DC: Department of Health and Human Services Office of the Surgeon General.

UNICEF. 2017. *State of the World's Children 2017*. New York: Communications Division UNICEF.

UNICEF. 2019a. *State of the World's Children 2019*. New York: Communications Division UNICEF.

UNICEF. 2019b. *Levels & Trends in Child Mortality 2019 Estimates Developed by the UN Inter-Agency Group for Child Mortality Estimation*. New York: UNICEF.

United Nations. 2019. *The Sustainable Development Goals Report 2019*. New York: UN.

United Nations Children's Fund and United Nations Special Rapporteur on the Right to Food. 2018. *Protecting Children's Right to a Healthy Food Environment*. Geneva: UNICEF and United Nations Human Rights Council.

United Nations Economic and Social Council. 2019. *Progress towards the Sustainable Development Goals Report of the Secretary-General E/2019/68*. New York: United Nations.

United Nations Office for Disaster Risk Reduction. 2015. *Disaster Risk Reduction and Resilience in the 2030 Agenda for Sustainable Development*. New York: UNISDR, Available online: https://www.preventionweb.net/files/46052_disasterriskreductioninthe2030agend.pdf (accessed on 25 September 2019).

Victora, Cesar G., Bernardo Lessa Horta, Christian Loretde Mola, Luciana Quevedo, Ricardo Tavares Pinheiro, Denise P. Gigante, Helen Gonçalves, and Fernando C. Barros. 2015. Association between breastfeeding and intelligence, educational attainment, and income at 30 years of age: A prospective birth cohort study from Brazil. *Lancet Globle Health* 3: e199–e205. [CrossRef]

Victora, Cesar G., Rajiv Bahl, Aluísio J. D. Barros, Giovanny V. A. Franca, Susan Horton, Julia Krasevec, Simon Murch, Mari Jeeva Sankar, Neff Walker, Nigel C. Rollins;, and et al. 2016. Breastfeeding in the 21st century: Epidemiology, mechanisms, and lifelong effect. *The Lancet* 387: 475–90. [CrossRef]

Walker, Susan P., Theodore D. Wachs, Sally Grantham-McGregor, Maureen M. Black, Charles A. Nelson, Sandra L. Huffman, Helen Baker-Henningham, Susan. M. Chang, Jena. D. Hamadani, Betsy Lozoff, and et al. 2011. Child Development 1 Inequality in early childhood: risk and protective factors for early child development. *The Lancet* 378: 1325–38. [CrossRef]

Walker, Christa L., Igor Rudan, Li Liu, Harish Nair, Evropi Theodoratou, Zulfiqar A. Bhutta, Katherine L. OBrien ', Harry Campbell, and Robert E. Black. 2013. Global burden of childhood pneumonia and diarrhoea. *The Lancet* 381: 1405–16. [CrossRef]

Walters, Dylan D., Linh T. H. Phan, and Roger Mathisen. 2019. The cost of not breastfeeding: global results from a new tool. *Health Policy and Planning* 34: 407–17. [CrossRef] [PubMed]

White, Becky, Roslyn C. Giglia, James A. White, Satvinder Dhaliwal, Sharyn K. Burns, and Jane A. Scott. 2019. Gamifying Breastfeeding for Fathers: Process Evaluation of the Milk Man Mobile App. *JMIR Pediatr Parent* 2: e12157. [CrossRef]

Whitehead, Richard. 2020. Dairy Majors investing in Vietnam. Available online: https://www.dairyreporter.com/Article/2020/01/08/Dairy-majors-investing-as-Vietnam-ramps-up-milk-production (accessed on 25 May 2020).

WHO. 1981. *International Code of Marketing of Breast-milk Substitutes*. Geneva: WHO.

WHO. 2019. *Protecting, Promoting and Supporting Breastfeeding in Facilities Providing Maternity and Newborn Services*. Geneva: World Health Organization, Available online: https://www.who.int/elena/titles/full_recommendations/breastfeeding-support/en/ (accessed on 5 June 2020).

WHO. 2020a. *Marketing of Breast-Milk Substitutes: National Implementation of the International Code, Status Report 2020*. Geneva: World Health Organization.

WHO. 2020b. *The International Code of Marketing of Breast-Milk Substitutes: Frequently Asked Questions on the Roles and Responsibilities of Health Workers*. Geneva: World Health Organization.

WHO and UNICEF. 2020. *WHO Policy Brief on International Trade Agreements and Implementation of the International Code of Marketing of Breast-Milk Substitutes*. Geneva: World Health Organization and the United Nations Children's Fund.

WHO Collaborative Group. 2000. Effect of breastfeeding on infant and child mortality due to infectious diseases in less developed countries: A pooled analysis. WHO Collaborative Study Team on the Role of Breastfeeding on the Prevention of Infant Mortality. *The Lancet* 355: 451–55. [CrossRef]

WHO UNICEF. 2018. *Implementation Guidance: Protecting, Promoting and Supporting Breastfeeding in Facilities Providing Maternity and Newborn Services—the Revised Baby Friendly Hospital Initiative. Licence: CC BY-NC-SA 3.0 IGO*. Geneva: Geneva: World Health Organization.

Wise, Jacqui. 2017. Plain cigarette packs may cut number of smokers, evidence review finds. *BMJ* 357: j2068. [CrossRef]

World Health Organization. 2009. *Acceptable Medical Reasons for Use of Breastmilk Substitutes*. Geneva: WHO, Available online: https://www.who.int/nutrition/publications/infantfeeding/WHO_NMH_NHD_09.01/en/ (accessed on 29 November 2019).

World Health Organization. 2011. *Exclusive Breastfeeding for Six Months Best for Babies Everywhere*. Geneva: WHO, Available online: https://www.who.int/mediacentre/news/statements/2011/breastfeeding_20110115/en/ (accessed on 21 September 2019).

World Health Organization. 2017. *The Goals within a Goal: Health Targets for SDG 3.* Geneva: WHO, Available online: https://www.who.int/sdg/targets/en/ (accessed on 16 September 2019).

World Health Organization. 2018a. Family Planning and Contraception. Available online: http s://www.who.int/news-room/fact-sheets/detail/family-planning-contraception (accessed on 4 October 2019).

World Health Organization. 2018b. *International Code of Marketing of Breast-Milk Substitutes 1981 with Updates to 2018.* Geneva: WHO, Available online: https://www.who.int/nutritio n/netcode/resolutions/en/ (accessed on 8 September 2020).

World Health Organization. 2019a. *Trends in Maternal mortality 2000 to 2017: Estimates by WHO, UNICEF, UNFPA, World Bank Group and the United Nations Population Division.* Geneva: WHO.

World Health Organization. 2019b. *Infant Feeding for the Prevention of Mother-to-Child Transmission of HIV.* Geneva: WHO, Available online: https://www.who.int/elena/ti tles/hiv_infant_feeding/en/ (accessed on 24 September 2019).

World Health Organization. 2020. *Breastfeeding and COVID-19 Scientific Brief.* Geneva: WHO, Available online: https://www.who.int/news-room/commentaries/detail/breastfeeding-and-covid-19 (accessed on 11 July 2020).

Xin, Hao, and Richard Stone. 2008. TAINTED MILK SCANDAL Chinese Probe Unmasks High-Tech Adulteration With Melamine. *Science* 322: 1310–11. [CrossRef]

Xu, Fenglian, Zhuoyang Li, Colin Binns, Michelle Bonello, Marie-Paule Austin, and Elizabeth Sullivan. 2014. Does infant feeding method impact on maternal mental health? *Breastfeeding Medicine* 9: 215–21. [CrossRef]

Yusuff, Aza Sherin, Li Tang, Colin W. Binns, and Andy H. Lee. 2015. Prevalence and risk factors for postnatal depression in Sabah, Malaysia: A cohort study. *Women and Birth* 28: 25–29. [CrossRef]

Zhang, Min, Xi Xie, Andy H. Lee, and Colin W Binns. 2004. Prolonged lactation reduces ovarian cancer risk in Chinese women. *European Journal of Cancer Prevention* 13: 499–502. [CrossRef]

Taxation of Tobacco, Alcohol and Sugar-Sweetened Beverages for Achieving the Sustainable Development Goals

Violeta Vulovic and Frank J. Chaloupka

1. Introduction

NCDs cause more than two-thirds of mortality in the world. More than 40 percent of these deaths are premature, and most of them are preventable. About 80 percent of NCD attributable premature deaths happen in low- and middle-income countries (LMICs). Moreover, low- and middle-income households bear 67 percent of NCD disease burden (IHME 2019), and one third of these deaths affects the economically productive age group, the age group below 70. Numerous studies show that tobacco use, obesity, and diabetes, especially among lower socioeconomic groups, have been high risk factors for NCDs (Niessen et al. 2018), and can lead to long-term economic consequences for individuals, their families, and the society. Around 60 percent of deaths from HCDs are caused by cardiovascular and respiratory diseases and diabetes. The number of deaths from these diseases has increased globally by 21.7 percent between 2000 and 2015. In terms of their share in total NCDs, there has been a very modest decline, from 61 percent in 2000 to 59 percent in 2015, primarily driven by a decline in high-income countries (HICs) (Figure 1). Potential economic implications of NCDs can be enormous if this trend continues, with an estimated cumulative global output loss by 2030 of USD 47 trillion (Bloom et al. 2012).

SDGs represent a global plan of action to end poverty, fight inequality, tackle climate change, and ensure that people live in peace and prosperity by 2030 (UN 2015b). Among its 17 goals, SDG target 3.4 under SGD 3 (good health and well-being for people) focuses on reducing premature mortality attributed to NCDs by one third by 2030. Progress on SDG target 3.4 is of a major relevance in accomplishing at least five other SGDs, specifically, SDG 1 (reducing poverty), SDG 2 (zero hunger), SDG 4 (quality education), SDG 8 (decent work and economic growth), and SDG 10 (reduced inequalities).

NCDs tend to be disproportionally clustered in lower socioeconomic groups of a society. While evidence on the relationship between the socioeconomic status and NCDs in HICs is extensive, there is little systematic evidence to support this link in LMICs. There are a few ways of how NCDs and poverty (SDG 1) are related. Firstly, poverty is linked to limited access to health care, and to an unhealthy diet and use of tobacco, which are targeted under SDG 3 (The Lancet 2017). People living in poverty are less able to practice healthy living and are more predisposed to suffering from chronic diseases (Wagstaff 2002).

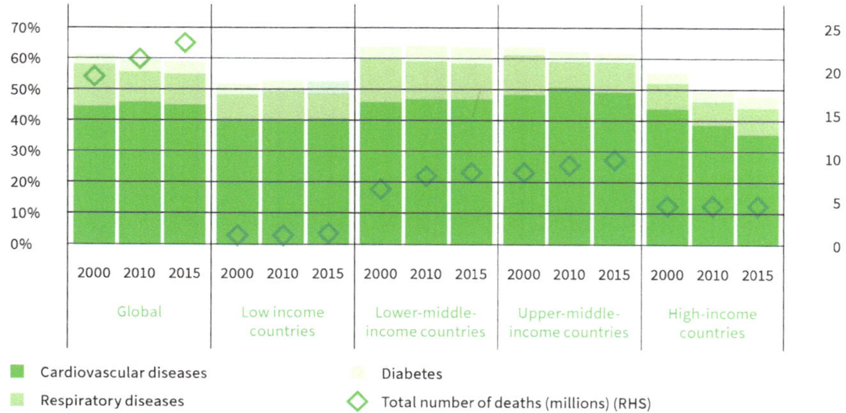

Figure 1. Deaths from selected NCDs by country-income group. Deaths caused by cardiovascular diseases, respiratory diseases, and diabetes total presented as % in total NCDs (LHS); total number of deaths from these three diseases presented in millions (RHS). Source: Authors' calculations using WHO (2018a).

Good nutrition is estimated as a necessary determinant for reducing NCDs and meeting most of SDGs (WHO 2017). Malnutrition at an early stage in life increases the risk of developing chronic diseases and hinders children's cognitive development and growth (Fall et al. 2010; Boney et al. 2005). Moreover, low socioeconomic status increases one's risk to develop chronic diseases and predisposes them and their families to economic hardship due to high out-of-pocket costs, lost earnings and employment, and then impoverishment (Hanratty et al. 2007). In such situations, families can lose the ability to support opportunities for education and human development of their children, especially when they have to stay at home and provide care to ill family members (Engelgau et al. 2012). Even those who are insured can incur very high medical costs of treatment of NCDs when they have low income or limited coverage (Jan et al. 2018), and they are commonly not able to pay for it (Levesque et al. 2007).

Better economic conditions and quality education enhance the health outcomes of household members, while low socioeconomic status leads to poor health, which further reduces their income status and may lead to poverty (Nugent et al. 2018; Jan et al. 2018). High premature mortality and morbidity from NCDs reduce workers' labor productivity and earnings and their family income (Bertram et al. 2018) and hinders a country's economic growth and development (SDG 8). Without a reduction in NCD attributed mortality and an increase in economic growth, especially in LMICs, the world will see not only the continued but deepening inequality within and between countries (SDG 10).

Around 39 percent of 40 million NCD attributed deaths in 2015 was caused by tobacco use (8 million), heavy alcohol consumption (3 million), and obesity (4.5 million) (WHO 2018b, 2019c; Gakidou et al. 2017). One of the major causes of obesity, globally, has been an increase in consumption of SSBs. Taxation is one of the most cost-effective measures that encourage healthy behavior, and prevent and reduce NCDs. By increasing prices, they can discourage consumption of tobacco, alcohol, and SSBs, and generate additional government revenues much needed for development spending (NCI and WHO 2016; Wagenaar et al. 2010; Sornpaisarn et al. 2013; Thow et al. 2018). However, despite their potential, these taxes have been frequently opposed and underused by the policymakers. Current tobacco tax collection in most LMICs is still well below 1 percent of GDP, which does not come even close to covering the estimated healthcare costs associated with tobacco-related illnesses estimated at well above 1.8 percent of GDP (Goodchild et al. 2018).

One of the most commonly used arguments by the policymakers for opposing taxation of tobacco, alcohol, and SSBs is that they may be regressive. While the evidence from developing countries on the distributional impact of taxation of tobacco, alcohol, and SSBs is limited, the findings from available studies suggest that the concern about their harmful effect on the poor may be overstated and that, on the contrary, they can result in health and welfare gains without imposing an excessive burden on the poor (Sassi et al. 2018).

This paper reviews existing evidence on adverse effects of consumption of tobacco, alcohol, and SSBs on health (SDG 3), and their relevance for achieving a few other related SDGs, including poverty reduction (SDG 1), nutrition (SDG 2), education (SDG 4), economic growth (SDG 8), and reduced inequality (SDG 10). Based on this evidence and evidence on the effectiveness of taxation in influencing consumer behavior and in collecting additional much needed revenues, this paper also builds a case for accelerating tax policy to meet development goals. The rest of the paper is organized as follows. Section 2 reviews the evidence on the health and development impacts of consumption of tobacco, alcohol, and SSBs. Section 3 discusses the evidence on impact of tax and price policy on consumption of these products and revenue impact of policy changes. Finally, Section 4 concludes the paper.

2. Impact of Consumption of Tobacco, Alcohol and SSBs on Selected SDGs

2.1. Tobacco

There is a significant body of research offering evidence on a negative impact of tobacco consumption on health and economic development. It has been estimated that tobacco causes more than 8 million deaths per year (WHO 2019c). Most of those deaths (7 million) are attributed to direct tobacco use, and more than 1 million to the exposure to second-hand tobacco smoke. Around 80 percent of these deaths happen in LMICs, and most of them are preventable. In terms of the regional disparity, almost 60 percent of tobacco-attributable deaths happen in South-East Asia and Western Pacific regions (Figure 2). Around half of lifetime smokers die before reaching 70, while quitting before the age of 40 reduces the mortality risk by 90 percent (Jha et al. 2015). Tobacco use significantly increases the risk of death from lung and other cancers, heart disease, stroke, respiratory disease, and tuberculosis (WHO 2019c).

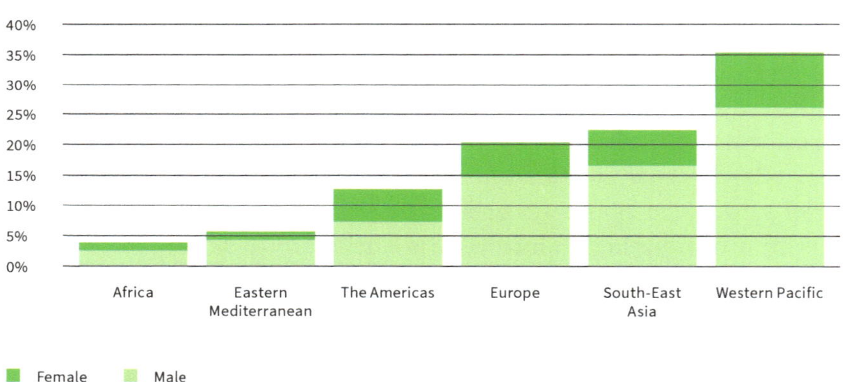

Figure 2. Smoking-attributable deaths by gender and WHO region, 2016 (% in total by gender). Source: Authors' calculations using GBD 2015 Risk Factors Collaborators (2016).

The evidence on the impact of tobacco use on health outcomes in HICs has been well documented, and it shows significant tobacco-related health disparities (TRHDs) among lower socioeconomic groups (Haustein 2006). This trend is in line with disparities in cigarette consumption showing relatively higher consumption by lower-income groups (Figure 3). Moreover, this disparity has been increasing, as smoking prevalence has declined relatively more among higher income groups than among those below the poverty line (NCI and WHO 2016). On the other hand, while still limited, the evidence from LMICs has been increasing, and showing a similar trend for male smokers, as lower income men are significantly more likely to smoke than the higher income men (Efroymson et al. 2001). However, the picture is different among women. While in about two thirds of countries the poorest women are significantly more likely to smoke than the wealthiest women, in about one third of countries, mostly in Eastern Europe, the case is the opposite (Hosseinpoor et al. 2012).

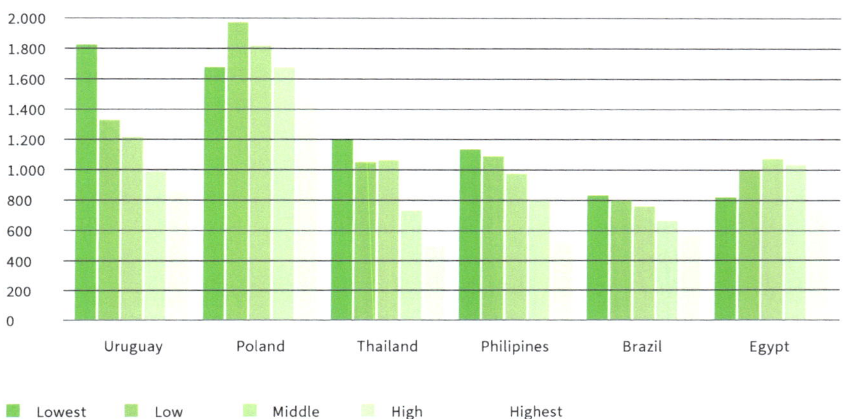

Figure 3. Cigarette consumption by income group in selected countries in 2009. Measured by an average number of cigarette sticks per person. Source: Authors' calculations using CDC (2019).

The economic impact of tobacco use can be very high. Evidence suggests that total global economic cost of smoking in 2012 was around 1.8 percent of global GDP (PPP$) (Figure 4), of which 25 percent were direct medical costs, and 75 percent indirect costs due to morbidity and mortality (Goodchild et al. 2018). LMICs accounted for almost 40 percent of this cost, while regionally, most of the cost (70 percent) was in the Americas and Europe. A systematic review of studies from various countries conducted between 1990 and 2011 finds that estimates vary greatly, depending on the scope, type of data, and methodology (NCI and WHO 2016). In LMICs, estimated direct and indirect costs of tobacco use range from 0.1 percent of GDP in Lao PDR to 3.4 percent of GDP in Philippines, while in HICs it is between 0.3 percent of GDP and 2 percent of GDP.

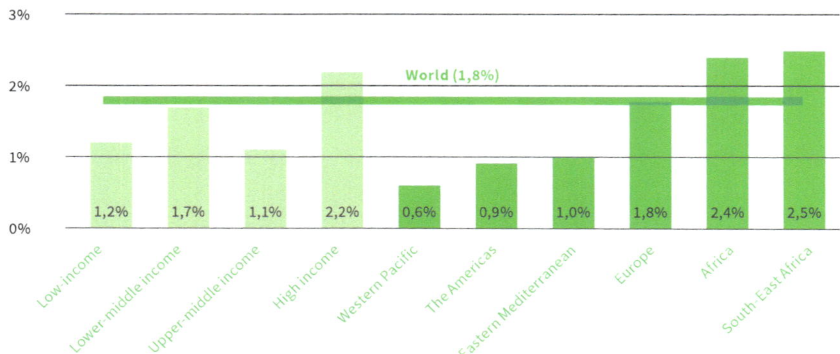

Figure 4. Economic costs of smoking by country-income group and WHO region, 2012 (% of GDP). Source: Authors' calculations using Goodchild et al. (2018).

Through its adverse effect on physical health, tobacco use can significantly negatively impact people's financial health and socioeconomic status, even in the long run. If a household's principal earner is a smoker and develops a disease, this would lead to medical costs associated to a treatment of that illness, and possibly a loss of earnings due to disability, and even mortality. In 1998, in China, 30.5 million people in urban areas, and 23.7 million people in rural areas were impoverished due to direct tobacco spending and medical costs for treatment of tobacco-attributable illnesses (Liu et al. 2006). As workers from poorer households commonly have lower level of education and lower skills, and usually work in labor intensive jobs, their disability often results in a relatively greater loss of earnings than with people with higher income and more resources.

Tobacco use can contribute to or exacerbate both primary and secondary poverty (NCI and WHO 2016; John et al. 2011). Primary poverty is defined as a situation in which an individual has no sufficient resources to afford necessities of life, while secondary poverty refers to a situation in which these resources are sufficient, but are not used effectively (Liu et al. 2006). There are two main tobacco-associated factors contributing to the impoverishment, the opportunity cost of tobacco use, and the burden due to illness, disability, and mortality.

With limited resources, spending on tobacco products requires a trade-off in consumption and crowds out basic necessities—food and non-food items (clothing, housing, education, healthcare, fuel). In India, spending on tobacco reduces spending on food, education, and entertainment (John 2008). In Bangladesh, it crowds out spending on education and healthcare (Do and Bautista 2015), while in South Africa, low income households sacrifise spending on dairy, fruits, nuts, and oils due to spending on tobacco (Chelwa and Koch 2019). In other words, this foregone consumption of basic necessities is called the opportunity cost of tobacco use. Tobacco use can hamper the standards of living of all income groups, but relatively more with poorer than with richer households (Pu et al. 2008). These changes in the household's internal resource allocation could lead to a poor diet and malnutrition in children, poor health, and decreased opportunities to obtain education and gain skills (Semba et al. 2007; Paraje and Araya 2018; Virk-Baker et al. 2019).

Moreover, numerous studies provide evidence on the wage-gap between smokers and non-smokers (Hotchkiss and Pitts 2013; Bondzie 2016). The estimated gap ranges from as little as 1.5–3.5 percent to more than 20 percent, depending on which factors that may contribute to lower earnings have been accounted for, in addition to tobacco use (Leigh and Berger 1989; Hotchkiss and Pitts 2013). In an attempt to explain the wage-gap due to tobacco use, studies have tried to evaluate the underlying causes, and have identified different reasons, from deteriorated health to lower productivity. In addition, it has been found that smokers place a higher emphasis on gratification in the present over their future well-being, and therefore tend to make relatively lower investments in education and a healthy life-style than non-smokers.

Economic costs from lost productivity resulting from presenteeism, absenteeism from work, and premature death of employees due to tobacco-related illnesses can be substantial, and consist of lost earnings for employees and lost revenues for the employers. Most of the available evidence comes from the HICs. It has been estimated that smokers are absent from work 6.6 more days per year than non-smokers in the US (Bunn et al. 2006) and 2.7 more days in the UK (Weng et al. 2013). In terms of lost productivity, the estimated cost of smokers in the US is around $US 151 billion (0.9 percent of GDP) and around $US 6 billion (0.03 percent of GDP) for non-smokers due to second-hand smoke (SHS) exposure (USDHHS 2014), and in Australia around $AU 8 billion (or 0.9 percent of GDP) (Collins and Lapsley 2008). Moreover, the cost of presenteeism represents more than 50 percent of total lost productivity attributed to tobacco use in the US (Bunn et al. 2006). Evidence from the US, European Union, China and Japan shows that around four years post-cessation, recent quitters report already higher work productivity than current smokers (Baker et al. 2017; Suwa et al. 2017).

In a long term, due to insufficient resources to meet basic needs caused by tobacco use and burden due to tobacco-attributed illnesses, mortality, and morbidity, and resulting poor nutrition and health, and poorer education and skills, a household can end up in a poverty trap (Haustein 2006; Liu et al. 2006), which may impair a country's welfare and economic development, and would widen social inequality.

2.2. Alcohol

There is an extensive body of research on the health risks of alcohol consumption. The relevance of alcohol consumption to health and global development is reflected in its inclusion within the SDGs. Specifically, SDG 3.5 focuses on strengthening prevention and treatment of substance abuse, including harmful use of alcohol.

Harmful use of alcohol contributed to 3 million deaths globally (WHO 2018b), which was about 39 deaths per 100,000 people (Figure 3), or 5.3 percent of all global deaths in 2016 (Figure 5). There is a significant variation in number and percentage of alcohol-attributable deaths by country-income group and WHO region. In 2016, the burden of deaths was the highest in lower-middle income countries (46.2 deaths per 100,000 people) and low-income countries (42.1 deaths per 100,000 people) and, regionally, in Africa (70.6 deaths per 100,000 people) and Europe (62.8 deaths per 100,000 people) (Figure 6). While there is no safe level of tobacco use, harm caused by alcohol use depends on the volume and patterns of drinking. In 2010, alcohol-attributable cancer contributed to 4.2 percent of all cancer deaths (Parry et al. 2011). It has also been shown that alcohol increases the risk of coronary heart disease and stroke (Corrao et al. 2004), and inadequate nutrition and obesity (Traversy and Chaput 2015).

As in case of tobacco, economic costs of alcohol use can be very high, and range substantially between countries. Global evidence, which, until recently, has mostly been from HICs, on estimated economic burden of alcohol misuse ranges from 0.45 to 5.44 percent of GDP (Thavorncharoensap et al. 2009). In the US, the estimated economic cost of excessive drinking in 2006 was US$ 223.5 billion (1.7 percent of GDP), of which 72.2 percent was from lost productivity and 11.0 percent from healthcare costs (Bouchery et al. 2011). Similar, the estimated cost in the EU in 2003 was 1.2 percent of GDP (Anderson and Baumberg 2006), and in the UK in 2009 was 1.4 percent of GDP (HM Government 2012). In Sri Lanka, it was estimated that the economic cost of alcohol-related illnesses in 2015 was US$ 885.9 million, or 1.1 percent of GDP (Ranaweera et al. 2018). A more comprehensive study conducted in South Africa included both tangible and intangible costs, and estimated total economic cost of harmful use of alcohol in 2009 at 10–12 percent of GDP (WHO 2014).

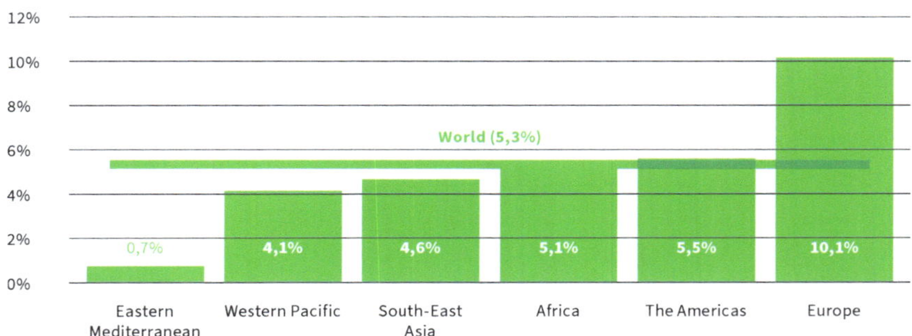

Figure 5. Alcohol-attributable deaths (% of all deaths) in 2016 by WHO region. Source: Authors' calculations using WHO (2019b).

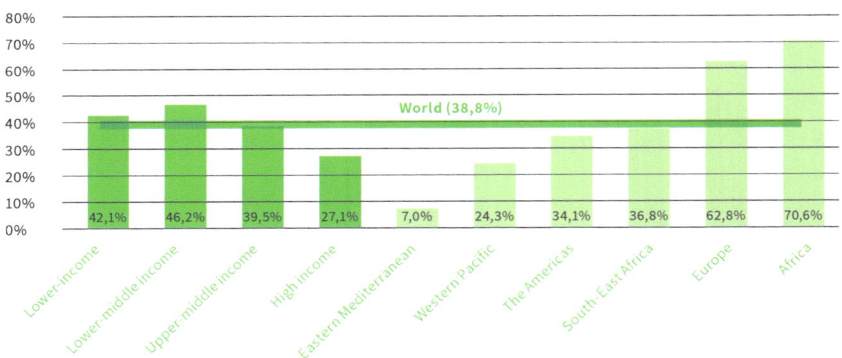

Figure 6. Alcohol-attributable deaths per 100,000 people (age-standardized) in 2016, by country-income group and WHO region. Source: Authors' calculations using WHO (2018b).

Alcohol consumption has multiple social and economic impacts. Firstly, harmful alcohol use and poverty are connected in many ways, as alcohol use can be both a response to, and a driver of, poverty. The risk of poverty is not related only to direct spending on purchase of alcohol, but also the medical cost for treatment of related illnesses and lost earnings due to disability and mortality. Evidence shows that disadvantaged socioeconomic groups are more likely to bear heavier alcohol-related burdens than the higher income groups with the same level of consumption (Smith and Foster 2014). In LMICs, as in case of tobacco, alcohol use tends to crowd out spending on basic necessities, such as healthy food, education, and healthcare. In Sri Lanka, the poorest users of alcohol and tobacco spend more than 40 percent of their income on these items (De Silva et al. 2011). In Madagascar, alcohol consuming households spent 14.1 percent of average monthly income on alcohol in 2010, with

33.6 percent of these households declaring that they had never been able to save (Tsikomia and Şarpe 2012). Even in countries with relatively low spending on alcohol and tobacco, such as Taiwan, the crowding out effect can be damaging to the lowest income households (Pu et al. 2008).

In many vulnerable socioeconomic groups, children bear a disproportionate burden of household's alcohol consumption. With limited resources, spending on alcohol crowds out spending on food and education, depriving children of nutrients and their right to primary education, both necessary for living a productive life. Moreover, heavy alcohol use among youth has been found to diminish chances of completing a degree, especially among male youth from lower socioeconomic background (Staff et al. 2008), and chances of being employed after college graduation and of securing employment (Bamberger et al. 2018).

Work problems associated with alcohol use are very relevant in many market economies due to their high costs of alcohol-related productivity losses and other work problems (Rehm and Rossow 2001). The relationship between alcohol use and earnings has received significant attention in the literature, especially in HICs. The empirical evidence on the impact of alcohol abuse on earnings is inconclusive. While some earlier studies suggested that moderate drinking may even be beneficial for work performance (Baum-Baicker 1985), but that the wage premium may decline as alcohol use increases (Berger and Leigh 1988), some more recent studies offer a slightly different insight. A study from Canada found that heavy alcohol users have lower earnings and lower returns to higher education relative to moderate or abstinent users (Hamilton and Hamilton 1997). Similarly, in the US, evidence suggests that alcoholism leads to lower earnings, but more due to reduced number of work hours that to a reduction in hourly wage (Renna 2008).

2.3. Sugar-Sweetened Beverages

Obesity is a form of malnutrition and represents a significant risk to public health and economic development. In 2016, 39 percent of world's adult population (18 years and above) were overweight and 13 percent were obese, which was an increase from 2000, with 30.8 and 8.7 percent, respectively (Figure 7). In recent years, the increases have been happening among children and young adults as well. In 2016, around 41 million children under the age of 5, and 340 million children and adolescents aged 5–19, were either overweight or obese. Global trends show a positive relationship between wealth and obesity (Low et al. 2009; Misra and Khurana 2008), as almost 60 percent of population in upper-middle-income and more than 80 percent in high-income countries were either overweight or obese in 2016. However, lower income countries have been catching up, as they have been experiencing a much higher increase than higher income countries, with more than 50 percent growth in overweight and obesity from 2000 to 2016 (Figure 7). While HICs do have a relatively higher obesity problem than LMICs, with economic development in LMICs, such as in China and India, obesity develops as well (James 2008). Almost half of overweight and obese children under the age of 5 now live in Asia, and the number has increased by almost 50 percent in Africa since 2000.

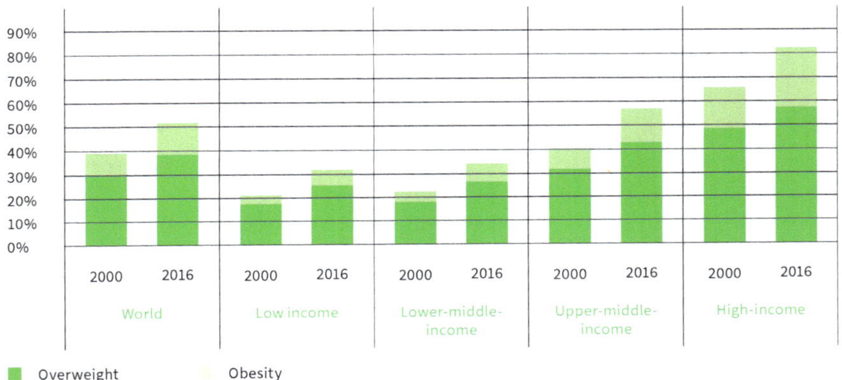

Figure 7. Prevalence of overweight and obesity among adults (18+ years of age) by country-income group, 2000 and 2016 (%). Source: Authors' calculations using WHO (2019a).

In 2016, 4.5 million deaths worldwide was attributed to obesity, in comparison to 3.5 million in 2006 (Gakidou et al. 2017). Diabetes caused 1.6 million deaths in 2015, which was 65.6 percent more than in 2000 (Figure 8). The increase has been much more rapid in LMICs than in HICs, and, by region, in Africa, Eastern Mediterranean, and South-East Asia.

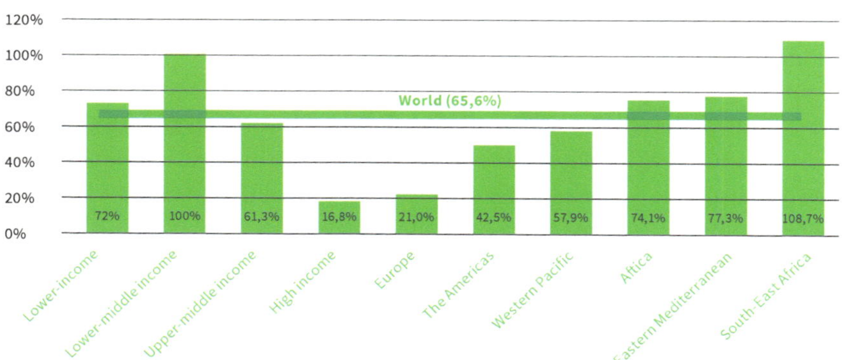

Figure 8. Percent increase in diabetes attributed mortality between 2000 and 2015, by country-income group and WHO region. Source: Authors' calculations using WHO (2018a).

Overweight and obesity are among the main contributors to poor health condition, as they may increase the risk of a large number of health problems, either independently or in association with other diseases (Kopelman 2007). Overweight and obese children are more likely to develop serious health problems, including type-2 diabetes and high blood pressure, but also the psychological disorders. Added sugar intake has been associated with multiple health risks, both among children and adults (Bovi et al. 2017). SSBs, in particular, are known to have a very high sugar content, and have been highly associated with overweight and obesity (Malik et al. 2006) and high blood pressure (Malik et al. 2014). The same amount of sugar and calories consumed in solid foods results in a significantly lower overall calorie intake than when consumed as liquids, because calories are less satiating in a liquid form (Mourao et al. 2007).

The economic costs attributed to obesity can be substantial and will only increase if the current trend continues. The global economic cost from obesity has been estimated at USD 2 trillion per year, which accounts for 1.5 percent of the 2018 global GDP (PPP$) (Dobbs et al. 2014). As the evidence from LMICs is still emerging, most evidence on economic costs of obesity is from the HICs. In the US, the direct medical cost of treatment of severe obesity-related illnesses in 2014 was estimated at US$ 69 billion (Wang et al. 2015), while the estimated cost of lost productivity due to absenteeism in 2012 was US$ 8.65 billion (Andreyeva et al. 2014). When converted to 2014 prices, these two costs together reached around 0.5 percent of national GDP. The evidence from Germany suggests significantly lower estimated economic costs of EUR 12.2 million in 2008 (Lehnert et al. 2015). A corresponding estimate for Thailand in 2009 was US$ 725.3 million, which accounted to 0.3 percent of its GDP (Pitayatienanan et al. 2014). Recent study estimated that, if current trends in overweight and obesity continue, around 92 million lives will be lost in OECD countries by 2050, and GDP will decline by 3.3 percent, on average (OECD 2019).

Different socioeconomic groups are differently impacted in HICs and LMICs by consumption of SSBs. A comparison of a relationship between obesity, economic development and socioeconomic status, including income and education level, in 67 countries, has produced some useful insights (Pampel et al. 2012). While obesity problem increases with a country's economic development, in HICs, higher socioeconomic groups are less likely to be obese, while in LMICs, those with higher income and education level are more likely to be obese. The authors highlight that this trend has a very relevant development implication. While economic development of a country is associated with an improvement in health condition of its citizens, it does not mitigate the problem of poor nutrition, as with development malnutrition is replaced by obesity. At the same time, some countries, such as India, with economic development struggle with both malnutrition and obesity (Ravishankar 2012).

3. Taxation of Tobacco, Alcohol and SSBs and SDGs

There is a very convincing body of evidence from various countries that substantially increasing prices of tobacco through taxation is the single most effective way to reduce tobacco use and alcohol consumption (NCI and WHO 2016; Anderson et al. 2009). Evidence on SSBs is still emerging and suggests that increased prices through sugar taxes are, similarly, an effective fiscal tool to promote health (Guerrero-López et al. 2017). Moreover, taxes on these three groups of products are also an effective policy measure for domestic revenue mobilization and meeting SDGs. However, despite their potential to mitigate the risk factors of consumption of tobacco, alcohol, and SSBs to health and NCDs, taxes on these products are underutilized by policymakers.

The economic justification for taxation of tobacco, alcohol, and SSBs is based on the principle of correcting negative externalities (Pigou 1920) by imposing a tax on a good, which would result in a reduction of consumption and increase in welfare. Moreover, imposing a tax corrects for the information failure in markets of these products, as consumers are misinformed about the full health consequences of their consumption (Chaloupka and Powell 2019). These costs, which consumers impose on themselves but do not correctly internalize, are called "internalities", and are used to make a distinction in the rationale behind "sin taxes" (Herrnstein et al. 1993; O'Donoghue and Rabin 2006).

While tobacco and alcohol can impose both direct externalities (e.g., second-hand smoke, disturbing others by making noise while drunk), and indirect externalities (e.g., medical costs for treating illnesses), the most important externalized costs of consumption of SSBs are indirect externalities. These financial costs associated with treatment of diseases attributed to tobacco, alcohol, and SSB consumption are, at least partly, financed by health insurance, and therefore born by the whole society. In other words, they are the costs of moral hazard (Allcott et al. 2019). Unlike externalities, internalities are born by the consumer. As explained above, they can happen either because of imperfect information or just because the consumer may be depreciating the potential future health costs associated to present consumption. Therefore, levying a tax on tobacco, alcohol, and SSBs would discourage consumption and reduce both negative externalities and internalities, and would increase welfare.

Extensive evidence on the responsiveness of consumption of addictive products to price changes over the last few decades showed that, in contrary to the conventional wisdom, the demand is in fact somewhat responsive in the short run, and more responsive in the long run (Chaloupka and Powell 2019). Most of the available evidence is on the impact of prices on consumption of tobacco products and alcohol, while the evidence on SSBs is still emerging.

Consumption of tobacco products in LMICs is most often more responsive to price than in HICs, with estimates from LMICs ranging between −0.2 and −0.8, clustering around −0.5, and in HICs clustering around −0.4. Evidence, although mostly from HICs, shows that youths are more responsive to tobacco price increases than adults. Economic theory suggests several reasons why this is the case, including limited income, peer effect, and lower addiction level due to a shorter period of consumption since initiation. A few studies have also estimated the impact of tobacco price increase on consumption by gender, but have produced mixed evidence. Moreover, several studies have examined the substitution among tobacco products, particularly among similar products (e.g., between cigarettes, roll-your-own tobacco, and little cigars) as a result of their relative price changes (NCI and WHO 2016). While results from HICs offer the evidence of substitution, the evidence from LMICs is mixed. Finally, as income increases, consumers tend to switch to more expensive products, such as from domestic to international brands (Chaloupka and Powell 2019).

In terms of the responsiveness of different socioeconomic groups, the evidence from both HICs and LMICs has been rather mixed. In the U.S. while more studies find that lower socioeconomic groups are more responsive to tobacco price change than higher socioeconomic groups (Farrelly et al. 2001; Chaloupka 1991; Colman and Remler 2008), some offer mixed evidence (Franks et al. 2007). In other HICs, such as UK and Australia, evidence also suggests that smokers in higher socioeconomic groups are much less impacted by price change than those in lower groups (Townsend et al. 1994; Siahpush et al. 2009). Findings from LMICs have been more mixed (Chaloupka et al. 2012; Levy et al. 2004). For example, while evidence from Bangladesh (Nargis et al. 2014), China (Verguet et al. 2015; Huang et al. 2015), or Indonesia (Adioetomo and Djutaharta 2005) shows significantly larger responses in lower income groups than among wealthier ones, findings from a few other countries, such as Nepal (Karki et al. 2003), Turkey (Onder 2002), or Thailand (Sarntisart 2003) have been less consistent.

While taxes on tobacco have been widely recognized as both a public health and a revenue generating tool, taxation of alcohol has been mostly used as a revenue instrument, but is increasingly gaining attention among policymakers as a tool for health promotion and disease prevention. There is a general consensus among scholars that increasing prices of alcohol may reduce consumption by youth and alcohol addiction (Chaloupka et al. 2002). Moreover, an increase in alcohol taxes may reduce motor vehicle accident mortality and suicide rates (Son and Topyan 2011). Evidence from HICs suggests that price elasticity ranges from -0.51 to -0.77, while limited studies from LMICs find it to be around -0.64 (Chaloupka and Powell 2019). Additionally, heavy drinkers are found to be less responsive, with price elasticity of around -0.28. Consumption of wine and spirits is more elastic (price elasticity between -0.68 and -0.80) than beer (between -0.36 and -0.46), and increase in income is associated with higher increase in alcohol consumption (Wagenaar et al. 2010; Gallet 2007). It has also been documented that alcohol users are more likely to smoke and that, in such cases, focusing on only one policy rather than a synergy would result in much smaller health benefits (Son and Topyan 2011; De Silva et al. 2011). Evidence on the different in response to change in alcohol price of different socioeconomic groups is missing.

Unlike taxation of tobacco and alcohol which have been levied in almost all countries, not as many governments levy a tax on SSBs. As of end of 2018, around 40 countries levy some type of tax on sugary drinks (WCRF 2018). Based on the existing evidence, taxation of SSBs leads to their lower consumption and to substitution to beverages with lower content of sugar (Backholer et al. 2018), with price elasticity clustering around -1.2 (Chaloupka and Powell 2019), and could reduce obesity rate (Escobar et al. 2013). Increasing prices on SSBs has a relatively larger positive impact on consumption in LMICs than in HICs. Evidence from Mexico, Ecuador, Chile, Guatemala, and South Africa finds price elasticity between -1.09 and -1.39 (Chaloupka and Powell 2019). Additionally, in almost all countries, consumption of bottom-income groups is more price elastic than of the higher-income groups (Sassi et al. 2018). Taxation of SSBs could have a substantial public health impact, while generating much needed revenues. A 2011 study from the US estimated that a national penny-per-ounce SSB tax would reduce consumption of SSBs by 24 percent within four years, and generate USD 79 billion of new tax revenues (Andreyeva et al. 2011). Mexico's SSB tax, which was introduced in 2014, reduced sales of SSBs by 5.5 percent in the first year, and by 9.7 percent in the second year after the implementation (Colchero et al. 2017). Similarly, it has been estimated that the newly adopted SSB tax in the Philippines could generate USD 813 million (0.25 percent of GDP) per year in additional revenues, and USD 627 million (0.19 percent of GDP) in healthcare savings (Saxena et al. 2019).

In addition to their public health objective, excise taxes are also an important source of government revenues. One of the central actions of the 2015 Addis Ababa Action Agenda is domestic revenue mobilization, which has recognized tobacco taxes as a key policy measure to reduce the burden of NCDs and help meet the SDGs (UN 2015a). This argument is supported by a great deal of evidence, including from LMICs, demonstrating that tobacco taxes are a powerful tool for reducing tobacco consumption while providing a source of government revenues (Chaloupka et al. 2012). For example, in South Africa, as a result of the tax increase, the real prices of cigarettes increased by 115 percent between 1993 and 2003, consumption decreased by around 40 percent, while real government revenues increased by almost 150 percent (Van Walbeek 2005). WHO has estimated that a 50 percent increase in taxes on cigarettes in LICs would generate enough additional revenues to finance more than 25 percent of their current government health spending (Stenberg et al. 2010).

Recent study commissioned by the Bloomberg Summers Task Force on Fiscal Policy for Health (Summan and Laxminarayan 2019) simulated health and revenue impacts of different increases of price of tobacco, alcohol, and SSB through tax increase, and found potential health and revenue benefits even in the short run. The study assessed what would happen over a period of 50 years if all countries increased taxes enough to raise prices by 20, 30, 40, and 50 percent (Table 1). The results showed the highest health impact of increased tobacco taxes, with around 11 million saved lives, with additional revenue of US$ 1.6 trillion in 2016 prices (around 1.3 percent of 2016 global GDP (PPP$)) if prices increased by only 20 percent. Potential revenue impact of alcohol taxes is the highest, mostly because current taxes are low. Alcohol price increase by 20 percent could generate additional US$ 8.9 trillion (in 2016 prices), or 7.3 percent of 2016 global GDP (PPP$). Potential impact of increases taxes on SSBs is relatively lower than on tobacco or alcohol, as consumption is not as widespread and only a set of products would be affected, and the estimate only captures the impact on body-mass index (BMI), but does not capture other effects, such as reduced obesity. Nevertheless, its potential impact is not negligible.

Table 1. Projected health and revenue impact of tax increase on tobacco, alcohol, and SSBs.

Price Increase via Tax	Tobacco			Alcohol			SSBs		
	Averted Deaths [1]	Gained Years of Life [1]	Tax Revenue Gain [2]	Averted Deaths [1]	Gained Years of Life [1]	Tax Revenue Gain [2]	Averted Deaths [1]	Gained Years of Life [1]	Tax Revenue Gain [2]
20%	10.8	212.0	1.6	9.4	238.7	8.9	0.8	23.7	0.7
30%	16.3	321.4	2.2	13.7	348.7	12.2	1.3	35.0	1.0
40%	21.8	428.6	2.6	17.9	455.0	14.8	1.7	46.5	1.2
50%	27.2	535.7	3.0	21.9	557.8	16.7	2.2	57.8	1.4

[1] In millions; [2] in trillions of 2016 USD. Source: The Task Force on Fiscal The Task Force on Fiscal Policy and Health (2019); Summan and Laxminarayan (2019).

One of the common concerns of policymakers against taxation of tobacco, alcohol, and SSBs is that they may be regressive, by mostly considering only the average ratio of the amount of tax paid and income. However, this measure does not take into account any health benefits of a tax from reduced consumption of unhealthy products. Lower income households most often respond to a price increase by reducing their consumption relatively more than the higher income households (Sassi et al. 2018). Moreover, as different socioeconomic groups can have different prevalence of consumption, the tax burden may be relatively higher in those groups with higher prevalence. As the goal of taxes on tobacco, alcohol, and SSBs is a reduction in prevalence and quantity of consumption, it is therefore important to account for a reduction in prevalence resulting from a tax when evaluating its distributional impact. In other words, in determining whether a tax is regressive or progressive, one needs to know the consumption patterns across income groups and their responsiveness to price (Summers 2018).

The distributional impact of these taxes depends on various factors, including prevalence rates and price elasticities by socioeconomic group, access to healthcare, and other fiscal policies. The potential equity implications of these taxes need to be addressed through other fiscal policy measures, such as direct subsidies or other targeted strategies that would incentivize behavioral change in vulnerable socioeconomic groups. Moreover, the revenue raised though taxes can be used for development spending. When they are used progressively or within a broader tax reform, they can relatively more benefit the poorer households. For example, the Philippines generated an additional US$ 1.5 billion from increased tobacco taxes between 2013 and 2015, and used most of this revenue to almost triple the national health insurance coverage for poor families (Nugent et al. 2018; Goodchild et al. 2017). Similarly, Denmark made changes in the income tax code to offset the potential equity impact of the new tax on saturated fats in 2011 (Sassi et al. 2018).

Furthermore, the burden of the tax should be put in the context of a burden of NCD disease (Summers 2018). Given that lower socioeconomic groups bear disproportionally larger burden of NCD disease (IHME 2019), and are, therefore, more likely to die prematurely, the burden of disease is regressive. Moreover, the economic costs associated with these diseases are also regressive, as lower income households are likely to bear relatively higher costs than the wealthier households. Therefore, an excise tax would, in fact, be progressive as the poor households would benefit relatively more through a significant reduction of these costs. For example, a 50 percent increase in cigarette excise taxes in China was estimated to reduce the tobacco-attributed medical costs borne by lower income household by US\$ 6.7 billion (Verguet et al. 2015). Similarly, if cigarette tax in Thailand was increased by 50 percent, it was estimated that, as a result of reduced consumption, the bottom-income group would bear only 6% of the new tax, and would benefit from 58 percent of averted premature deaths (Jha et al. 2012). In Finland, after the government reduced taxes on alcohol in 2003, alcohol-related mortality increased by 16 percent among men and 31 percent among women (Herttua et al. 2008).

Evidence from several countries shows that a tobacco tax is progressive because health benefits from reduced consumption due to tax and price increase more than offset the increase in tax liability, especially for lower socioeconomic groups (Fuchs et al. 2019). Another study compares the burden of excise taxes on tobacco and alcohol in Chile, Poland, and Turkey, and taxes on SSBs in a group of LMICs to assess their distributional impact (Sassi et al. 2018). The tax burden is found to be relatively larger for the bottom-income quintiles in case of tobacco, and for the top-income groups in case of alcohol and SSBs. However, when only households who consume these products are taken into account, the bottom-income groups bear the highest tax burden for all three products.

Indeed, in some countries tax structures are designed in a way to keep the prices of some products at a low level in order to keep them affordable for low-income consumers. While the argument for such a policy is that it is "pro-poor", in fact it is not as it leads to lower income groups bearing a disproportionate share of the health and economic burden of these products, due to their greater consumption (Bobak et al. 2000). As a result, such tax policy is indeed regressive. However, considering the empirical evidence presented above, even if the current tax is regressive, a tax increase can be progressive. To the extent that concerns about the impact on the poor persist, they can be addressed by using the additional revenues in ways that benefit the poor, such as expanding health coverage for lower-income households.

The Politics of Taxes on Tobacco, Alcohol, and SSBs: Selected Examples

Commonly, governments' reluctance to commit to a tax increase is due to a lack of evidence on the relevance of these taxes for health and development, low technical capacity of relevant government agencies, limited political commitment, and lack of coordination within the government. In addition, industry opposition and interference play a significant role. As a result, the legislative process is usually slow and often delayed. The industries are very fierce in their campaigns against a tax increase. They use their arguments strategically and present themselves as a very relevant stakeholder, with which they justify the need to be part of the policy dialogue. As a result, their interests often prevail against the public interests. However, it is not a random case that the proponents of the tax increase have successfully challenged the industry influence in the policy formulation (McCambridge et al. 2018). Some examples of successful policy reforms, despite very strong industry opposition, are presented below.

The Philippines had a long history of strong opposition from the tobacco industry to tax increases, and several politicians had close connections to the industry. Things changed in 2010 when Benigno Aquino III was elected as the new president, who, unlike his predecessors, did not accept the campaign contributions from the industry (Madore et al. 2015). Aquino had government reform and poverty reduction high on his political agenda, which included increasing sin taxes. Strong opposition from the industry was expected, and tobacco farmers and factory workers were mobilized to protest the bill. At the same time, most legislators in the Ways and Means Committee in the House of Representatives had known connections to the industry. As a result, the objectives of the proposed reform were strategically formulized around emphasizing that the additional revenues would be used for expanding the national health insurance (PhilHealth) coverage to low-income households and for economic development projects in tobacco growing provinces. Interestingly, empirical evidence on the economic and health impact of smoking, including estimates of the economic cost, already existed in the Philippines, but no policymaker used this evidence in any of the previous attempts to push for a tax increase (Madore et al. 2015).

Despite strong opposition, the bill was passed and signed into Law in 2012, becoming effective in January 2013. A compromise was made with the tobacco industry in terms of the tax rate simplification—instead of an immediate transition to a unitary system, a gradual move from the existing four-tier to a two-tier system was agreed on, with an implementation of the unitary system in 2017. Additionally, to gain the support for the reform of the alcohol tax, an agreement was made with San Miguel Brewery, which represented 90 percent of the Philippines beer market, to include its popular premium beer brand in the lower of the price tiers of the fermented liquor brands. While distilled spirits were taxed at a uniform rate starting 2013, the two-tier system for fermented alcohol transitioned to the uniform system in 2017. As a result, cigarette taxes increased by 340 percent between 2012 and 2013, resulting in 48 percent increase in the average price per cigarette pack. The collection of tobacco excise tax revenue in 2013 increased by 114 percent relative to 2012, while alcohol tax collections increased by 38 percent. Most of the additional funds (85 percent of tobacco tax and 100 percent of alcohol tax revenues) were earmarked for health, with 80 percent of the funds being used to increase the PhilHealth enrollment of low-income Filipinos, achieving 100 percent enrollment among this income group. It was reported that smoking prevalence among adults declined from 29 percent in 2012 to 26 percent in 2014, with a highest reduction in the lowest-income group (from 38 to 25 percent) and among adults between 18 and 24 years of age (from 35 to 18 percent) (Madore et al. 2015).

Similarly, Ukraine has a long history of tobacco industry price manipulation and interference in tobacco excise tax policy (Hoe et al. 2020). The goal of tobacco industry lobbying in Ukraine has, for a long time, been keeping tax and price increases at a low level to assure that the changes do not exceed the rate of inflation. However, despite their strong effort, for the first time since 1998, the increase in cigarette tax in 2008 finally outpaced inflation, and despite the 2009 economic recession, which reduced the real affordability of cigarettes by half, the government continued to substantially increase the tax until 2017. This reform was supported by newly elected president Petro Poroshenko, who signed the EU-Ukraine Association Agreement in 2014. As a result, during this nine-year period, the average excise tax increased 20 times and tobacco tax revenues increased 11 times, while smoking prevalence decreased by 35 percent and consumption of cigarettes by 46 percent. Moreover, in 2017, the government adopted a seven-year plan to further increase the specific excise tax by almost 30 percent in 2018 and annually by 20 percent between 2019 and 2024. In proceeding with the proposed reform, the government had three main concerns, including the impacts on revenue collection, the poor, and illicit trade. Extensive empirical evidence was produced and provided to the Ministry of Finance offering arguments and reassurance that these concerns were not justified.

As SSB taxes are relatively new, at least in comparison to taxation of tobacco and alcohol, not as many country examples are available to illustrate the political economy of the issue. That, however, does not mean that the resistance of the SSB industry has not been fierce. One good example is Mexico, where the SSB industry has long had ties with the Mexican government and society, including the former chief executive of Coca Cola Mexico being elected President of Mexico in 2010 (James et al. 2020). There had been attempts before 2012 to introduce the SSB tax in Mexico, but they were not successful. Opponents of the tax, including the SSB industry, sugar cane industry, and retailers used to run aggressive campaigns, calling on the Congress to oppose the tax, arguing that the tax would be regressive, would reduce employment, and would not reduce obesity rates. Finally, in 2012, the supporters, including civil society, politicians, and academics, effectively organized and supported the SSB tax proposal. The proposal also received strong support from the President Enrique Peña Nieto.

As the Ministry of Finance kept the tax plans secret, the opponents of the tax did not anticipate it and were not very active before the President announced the tax. One reason may be that they were relying on their historically strong ties with the government. Once it was sent to the Congress for consideration, the SSB industry began aggressively lobbying against the proposal. However, by the time the opponents managed to organize, the supporters already had messages framed around all negative implications of excessive SSB consumption and produced evidence providing counter-arguments to the anticipated industry arguments. As a result, the House of Deputies and the Senate passed the bill in October 2013 (James et al. 2020).

Scotland presents a case of implementing a minimum price per unit of alcohol, which was a successful policy reform, but not without a fierce opposition. The policy was proposed in 2012 with an objective of reducing consumption and its harmful impact. However, the Scotch Whiskey Association and the European trade associations for spirits and wines challenged it and took it to the EU Court of Justice. They argued that the policy would not be effective in meeting its objective, that it would hurt the poor, lead to illicit trade, and harm businesses (Shona et al. 2014). While the court did not dismiss the case, in December 2015 it requested additional evidence that an alternative policy, such as higher taxes, would not be effective in meeting public health objectives. Based on this evidence, the UK Supreme Court rejected the industry's case in November 2017, and the law came into effect on 1 May 2018.

Based on these examples, it is obvious that passing a tax on tobacco, alcohol, and SSBs is a difficult process and involves many factors. Factors that made a difference and provided support to the Ministry of Finance in proceeding with the reform were: high-level support and strong political will and collaboration between different parts of the Government working in partnership toward a common goal; thorough understanding of the political economy context in the country; available strong empirical evidence providing the counter-arguments to the concerns about the potential impact of the policy; and support provided by civil society and international organizations.

4. Conclusions

As it is becoming increasingly clear that economic prosperity and health are interdependent and that NCDs are a serious obstacle for achieving several SDG targets, fiscal policies targeting people's incentives for healthy behavior are becoming more and more appealing to the policymakers. When properly designed, these policies can improve health and raise much needed revenues, while eliminating the potential equity impact of taxes, thereby supporting the achievement of SDGs. NCDs are disproportionally clustered in lower socioeconomic groups of a society, and are a significant contributor to impoverishment, thereby worsening inequality. Tax policies can avert impoverishment, enhance workers productivity and economic growth by discouraging consumption of unhealthy products and improving diet, thereby contributing to prevention and control of NCDs and reducing the burden, and incentivizing support of human capital development.

Author Contributions: V.V. and F.J.C. conceptualized the manuscript. V.V. developed the first draft and F.J.C. provided inputs and edits. V.V. and F.J.C. finalized the draft. All authors have read and agreed to the published version of the manuscript.

Funding: This research received no external funding.

Conflicts of Interest: The authors declare no conflict of interest.

References

Adioetomo, Sri Moertiningsih, and Triasih Djutaharta. 2005. *Cigarette Consumption, Taxation, and Household Income: Indonesia Case Study.* Washington, DC: World Bank.

Allcott, Hunt, Benjamin Lockwood, and Dmitry Taubinsky. 2019. *Should We Tax Sugar-Sweetened Beverages? An Overview of Theory and Evidence.* Cambridge: National Bureau of Economic Research.

Anderson, Peter, and Ben Baumberg. 2006. Alcohol in Europe–Public Health Perspective: Report Summary. *Drugs: Education, Prevention and Policy* 13: 483–88. [CrossRef]

Anderson, Peter, Dan Chisholm, and Daniela C Fuhr. 2009. Effectiveness and Cost-Effectiveness of Policies and Programmes to Reduce the Harm Caused by Alcohol. *The Lancet* 373: 2234–46. [CrossRef]

Andreyeva, Tatiana, Frank J. Chaloupka, and Kelly D Brownell. 2011. Estimating the Potential of Taxes on Sugar-Sweetened Beverages to Reduce Consumption and Generate Revenue. *Preventive Medicine* 52: 413–16. [CrossRef] [PubMed]

Andreyeva, Tatiana, Joerg Luedicke, and Y. Claire Wang. 2014. State-Level Estimates of Obesity-Attributable Costs of Absenteeism. *Journal of Occupational and Environmental Medicine/American College of Occupational and Environmental Medicine* 56: 1120. [CrossRef]

Backholer, Kathryn, Stefanie Vandevijvere, Miranda Blake, and Marilyn Tseng. 2018. Sugar-Sweetened Beverage Taxes in 2018: A Year of Reflections and Consolidation. *Public Health Nutrition* 21: 3291–95. [CrossRef]

Baker, Christine L., Natalia M. Flores, Kelly H. Zou, Marianna Bruno, and Vannessa J. Harrison. 2017. Benefits of Quitting Smoking on Work Productivity and Activity Impairment in the United States, the European Union and China. *International Journal of Clinical Practice* 71: e12900. [CrossRef]

Bamberger, Peter A., Jaclyn Koopmann, Mo Wang, Mary Larimer, Inbal Nahum-Shani, Irene Geisner, and Samuel B. Bacharach. 2018. Does College Alcohol Consumption Impact Employment upon Graduation? Findings from a Prospective Study. *Journal of Applied Psychology* 103: 111. [CrossRef]

Baum-Baicker, Cynthia. 1985. The Psychological Benefits of Moderate Alcohol Consumption: A Review of the Literature. *Drug and Alcohol Dependence* 15: 305–22. [CrossRef]

Berger, Mark C., and J. Paul Leigh. 1988. The Effect of Alcohol Use on Wages. *Applied Economics* 20: 1343–51. [CrossRef]

Bertram, Melanie Y., Kim Sweeny, Jeremy A. Lauer, Daniel Chisholm, Peter Sheehan, Bruce Rasmussen, Senendra Raj Upreti, Lonim Prasai Dixit, Kenneth George, and Samuel Deane. 2018. Investing in Non-Communicable Diseases: An Estimation of the Return on Investment for Prevention and Treatment Services. *The Lancet* 391: 2071–78. [CrossRef]

Bloom, David E., Elizabeth Cafiero, Eva Jané-Llopis, Shafika Abrahams-Gessel, Lakshmi Reddy Bloom, Sana Fathima, Andrea B Feigl, Tom Gaziano, Ali Hamandi, and Mona Mowafi. 2012. *The Global Economic Burden of Noncommunicable Diseases*. PGDA Working Papers 8712. Cambridge: Program on the Global Demography of Aging.

Bobak, Martin, Prabhat Jha, Son Nguyen, and Martin Jarvis. 2000. Poverty and Smoking. In *Tobacco Control in Developing Countries*. Edited by Prabhat Jha and Frank J. Chaloupka. Oxford: Oxford University Press, pp. 41–62.

Bondzie, Eric. 2016. *Effect of Smoking and Other Economic Variables on Wages in the Euro Area*. MPRA Paper 69230. Munich: University Library of Munich.

Boney, Charlotte M., Anila Verma, Richard Tucker, and Betty R. Vohr. 2005. Metabolic Syndrome in Childhood: Association with Birth Weight, Maternal Obesity, and Gestational Diabetes Mellitus. *Pediatrics* 115: e290–6. [CrossRef] [PubMed]

Bouchery, Ellen E., Henrick J. Harwood, Jeffrey J. Sacks, Carol J. Simon, and Robert D. Brewer. 2011. Economic Costs of Excessive Alcohol Consumption in the U.S., 2006. *American Journal of Preventive Medicine* 41: 516–24. [CrossRef] [PubMed]

Bunn, William B., III, Gregg M. Stave, Kristen E. Downs, Jose Ma. J. Alvir, and Riad Dirani. 2006. Effect of Smoking Status on Productivity Loss. *Journal of Occupational and Environmental Medicine* 48. Available online: https://journals.lww.com/joem/Fulltext/2006/10000/Effect_of_Smoking_Status_on_Productivity_Loss.18.aspx (accessed on 23 October 2019). [CrossRef] [PubMed]

CDC (Center for Disease Control and Prevention). 2019. Global Tobacco Surveillance System Data (GTSSData). Available online: https://www.cdc.gov/tobacco/global/gtss/gtssdata/index.html (accessed on 26 October 2019).

Chaloupka, Frank J. 1991. Rational Addictive Behavior and Cigarette Smoking. *Journal of Political Economy* 99: 722–42. [CrossRef]

Chaloupka, Frank J., Michael Grossman, and Henry Saffer. 2002. The Effects of Price on Alcohol Consumption and Alcohol-Related Problems. *Alcohol Research & Health: The Journal of the National Institute on Alcohol Abuse and Alcoholism* 26: 22–34.

Chaloupka, Frank J., and Lisa M. Powell. 2019. *Using Fiscal Policy to Promote Health: Taxing Tobacco, Alcohol, and Sugary Beverages.* Backgroun Paper for the Task Force on Fiscal Policy and Health. New York: Bloomberg Philantropies. Available online: https://www.bbhub.io/dotorg/sites/2/2019/04/Using-Fiscal-Policy-to-Promote-Health-Taxing-Tobacco-Alcohol-and-Sugary-Beverages.pdf (accessed on 25 October 2019).

Chaloupka, Frank J., Ayda Yurekli, and Geoffrey T. Fong. 2012. Tobacco Taxes as a Tobacco Control Strategy. *Tobacco Control* 21: 172–80. [CrossRef]

Chelwa, Grieve, and Steven F. Koch. 2019. The Effect of Tobacco Expenditure on Expenditure Shares in South African Households: A Genetic Matching Approach. *PLoS ONE* 14: e0222000. [CrossRef]

Colchero, M. Arantxa, Juan Rivera-Dommarco, Barry M. Popkin, and Shu Wen Ng. 2017. In Mexico, Evidence of Sustained Consumer Response Two Years after Implementing a Sugar-Sweetened Beverage Tax. *Health Affairs* 36: 564–71. [CrossRef]

Collins, David J., and Helen M. Lapsley. 2008. *The Costs of Tobacco, Alcohol and Illicit Drug Abuse to Australian Society in 2004/2005.* Canberra: Commonwealth of Australia.

Colman, Gregory J., and Dahlia K. Remler. 2008. Vertical Equity Consequences of Very High Cigarette Tax Increases: If the Poor Are the Ones Smoking, How Could Cigarette Tax Increases Be Progressive? *Journal of Policy Analysis and Management* 27: 376–400. [CrossRef]

Corrao, Giovanni, Vincenzo Bagnardi, Antonella Zambon, and Carlo La Vecchia. 2004. A Meta-Analysis of Alcohol Consumption and the Risk of 15 Diseases. *Preventive Medicine* 38: 613–19. [CrossRef]

De Silva, Varuni, Diyanath Samarasinghe, and Raveen Hanwella. 2011. Association between Concurrent Alcohol and Tobacco Use and Poverty. *Drug and Alcohol Review* 30: 69–73. [CrossRef] [PubMed]

Bovi, Delli, Anna Pia, Laura Di Michele, Giuliana Laino, and Pietro Vajro. 2017. Obesity and Obesity Related Diseases, Sugar Consumption and Bad Oral Health: A Fatal Epidemic Mixtures: The Pediatric and Odontologist Point of View. *Translational Medicine @ UniSa* 16: 11–16.

Do, Young Kyung, and Mary Ann Bautista. 2015. Tobacco Use and Household Expenditures on Food, Education, and Healthcare in Low-and Middle-Income Countries: A Multilevel Analysis. *BMC Public Health* 15: 1098. [CrossRef] [PubMed]

Dobbs, Richard, Corinne Sawers, Fraser Thompson, James Manyika, Jonathan Woetzel, Peter Child, Sorcha McKenna, and Angela Spatharou. 2014. *Overcoming Obesity: An Initial Economic Analysis.* San Francisco: McKinsey Global Institute, Brussels: McKinsey & Company. Available online: https://www.mckinsey.com/~{}/media/McKinsey/Business%20Functions/Economic%20Studies%20TEMP/Our%20Insights/How%20the%20world%20could%20better%20fight%20obesity/MGI_Overcoming_obesity_Full_report.ashx (accessed on 8 November 2019).

Efroymson, Debra, Saifuddin Ahmed, Joy Townsend, Syed Mahbubul Alam, Amit Ranjan Dey, Ranjit Saha, Biplob Dhar, Aminul Islam Sujon, Kayum Uddin Ahmed, and Oliur Rahman. 2001. Hungry for Tobacco: An Analysis of the Economic Impact of Tobacco Consumption on the Poor in Bangladesh. *Tobacco Control* 10: 212. [CrossRef]

Engelgau, Michael M., Anup Karan, and Ajay Mahal. 2012. The Economic Impact of Non-Communicable Diseases on Households in India. *Globalization and Health* 8: 9. [CrossRef]

Escobar, Maria A. Cabrera, J. Lennert Veerman, Stephen M. Tollman, Melanie Y. Bertram, and Karen J. Hofman. 2013. Evidence That a Tax on Sugar Sweetened Beverages Reduces the Obesity Rate: A Meta-Analysis. *BMC Public Health* 13: 1072. [CrossRef]

Fall, Caroline H. D., Judith B. Borja, Clive Osmond, Linda Richter, Santosh K Bhargava, Reynaldo Martorell, Aryeh D. Stein, Fernando C. Barros, Cesar G. Victora, and The COHORTS Group. 2010. Infant-Feeding Patterns and Cardiovascular Risk Factors in Young Adulthood: Data from Five Cohorts in Low- and Middle-Income Countries. *International Journal of Epidemiology* 40: 47–62. [CrossRef]

Farrelly, Matthew C., Jeremy W. Bray, Terry Pechacek, and Trevor Woollery. 2001. Response by Adults to Increases in Cigarette Prices by Sociodemographic Characteristics. *Southern Economic Journal* 68: 156–65. [CrossRef]

Franks, Peter, Anthony F. Jerant, J. Paul Leigh, Dennis Lee, Alan Chiem, Ilene Lewis, and Sandy Lee. 2007. Cigarette Prices, Smoking, and the Poor: Implications of Recent Trends. *American Journal of Public Health* 97: 1873–77. [CrossRef]

Fuchs, Alan Tarlovksy, Patricio V Marquez, Sheila Dutta, and Fernanda Gonzalez Icaza. 2019. *Is Tobacco Taxation Regressive? Evidence on Public Health, Domestic Resource Mobilization, and Equity Improvements*. Washington, DC: World Bank.

Gakidou, Emmanuela, Ashkan Afshin, Amanuel Alemu Abajobir, Kalkidan Hassen Abate, Cristiana Abbafati, Kaja M. Abbas, Foad Abd-Allah, Abdishakur M. Abdulle, Semaw Ferede Abera, Victor Aboyans, and et al. 2017. Global, Regional, and National Comparative Risk Assessment of 84 Behavioural, Environmental and Occupational, and Metabolic Risks or Clusters of Risks, 1990–2016: A Systematic Analysis for the Global Burden of Disease Study 2016. *The Lancet* 390: 1345–422. [CrossRef]

Gallet, Craig A. 2007. The Demand for Alcohol: A Meta–analysis of Elasticities. *Australian Journal of Agricultural and Resource Economics* 51: 121–35. [CrossRef]

GBD 2015 Risk Factors Collaborators. 2016. Global, Regional, and National Comparative Risk Assessment of 79 Behavioural, Environmental and Occupational, and Metabolic Risks or Clusters of Risks, 1990–2015: A Systematic Analysis for the Global Burden of Disease Study 2015. *Lancet* 388: 1659–724.

Goodchild, Mark, Anne-Marie Perucic, Rose Zheng, Evan Blecher, and Jeremias Paul. 2017. The Health Impact of Raising Tobacco Taxes in Developing Countries. In *Tobacco Tax Reform—at the Crossroads of Health and Development*. Edited by Patricio V. Marquez and Blanca Moreno-Dodson. Washington, DC: World Bank Group. Available online: http://hdl.handle.net/10986/28494 (accessed on 23 October 2019).

Goodchild, Mark, Nigar Nargis, and Edouard Tursan d'Espaignet. 2018. Global Economic Cost of Smoking-Attributable Diseases. *Tobacco Control* 27: 58–64. [CrossRef] [PubMed]

Guerrero-López, Carlos M., Mishel Unar-Munguía, and M. Arantxa Colchero. 2017. Price Elasticity of the Demand for Soft Drinks, Other Sugar-Sweetened Beverages and Energy Dense Food in Chile. *BMC Public Health* 17: 180. [CrossRef]

Hamilton, Vivian, and Barton H. Hamilton. 1997. Alcohol and Earnings: Does Drinking Yield a Wage Premium? *The Canadian Journal of Economics/Revue Canadienne d'Economique* 30: 135–51. [CrossRef]

Hanratty, Barbara, Paula Holland, Ann Jacoby, and Margaret Whitehead. 2007. Review Article: Financial Stress and Strain Associated with Terminal Cancer—A Review of the Evidence. *Palliative Medicine* 21: 595–607. [CrossRef]

Haustein, Knut-Olaf. 2006. Smoking and Poverty. *European Journal of Cardiovascular Prevention & Rehabilitation* 13: 312–18. [CrossRef]

Herrnstein, R. J., George F. Loewenstein, Drazen Prelec, and William Vaughan Jr. 1993. Utility Maximization and Melioration: Internalities in Individual Choice. *Journal of Behavioral Decision Making* 6: 149–85. [CrossRef]

Herttua, Kimmo, Pia Mäkelä, and Pekka Martikainen. 2008. Changes in Alcohol-Related Mortality and Its Socioeconomic Differences after a Large Reduction in Alcohol Prices: A Natural Experiment Based on Register Data. *American Journal of Epidemiology* 168: 1110–18. [CrossRef]

Shona, Karen Wood, Chris Patterson, and Srinivasa Vittal Katikireddi. 2014. Implications for Alcohol Minimum Unit Pricing Advocacy: What Can We Learn for Public Health from UK Newsprint Coverage of Key Claim-Makers in the Policy Debate? *Social Science & Medicine* 102: 157–64.

HM Government. 2012. *The Government's Alcohol Strategy*; London: Her Majesty's Government in the United Kingdom. Available online: https://www.gov.uk/government/publications/alcohol-strategy (accessed on 8 November 2019).

Hoe, Connie H., Caitlin Weiger, and Joanna Cohen. 2020. Increasing Taxes on Tobacco: Advocacy Lessons from Ukraine. Johns Hopkins Bloomberg School of Public Health, Institute for Global Tobacco Control. Available online: https://globaltobaccocontrol.org/resources/increasing-taxes-tobacco-advocacy-lessons-ukraine (accessed on 4 August 2020).

Hosseinpoor, Ahmad Reza, Lucy Anne Parker, Edouard Tursan d'Espaignet, and Somnath Chatterji. 2012. Socioeconomic Inequality in Smoking in Low-Income and Middle-Income Countries: Results from the World Health Survey. *PLoS ONE* 7: e42843. [CrossRef] [PubMed]

Hotchkiss, Julie L., and M. Melinda Pitts. 2013. *Even One Is too Much: The Economic Consequences of Being a Smoker*. FRB Atlanta Working Paper Series 2013-3; Available online: https://papers.ssrn.com/sol3/papers.cfm?abstract_id=2359224 (accessed on 18 October 2019).

Huang, Jidong, Rong Zheng, Frank J. Chaloupka, Geoffrey T Fong, and Yuan Jiang. 2015. Differential Responsiveness to Cigarette Price by Education and Income among Adult Urban Chinese Smokers: Findings from the ITC China Survey. *Tobacco Control* 24 S3: iii76–iii82. [CrossRef] [PubMed]

IHME (Institute for Health Matrics and Evaluation). 2019. Global Health Data Exchange (GHDx). Available online: http://ghdx.healthdata.org/gbd-results-tool (accessed on 18 October 2019).

James, William Philip T. 2008. The Epidemiology of Obesity: The Size of the Problem. *Journal of Internal Medicine* 263: 336–52. [CrossRef] [PubMed]

James, Erin, Martín Lajous, and Michael R. Reich. 2020. The Politics of Taxes for Health: An Analysis of the Passage of the Sugar-Sweetened Beverage Tax in Mexico. *Health Systems & Reform* 6: e1669122.

Jan, Stephen, Tracey-Lea Laba, Beverley M. Essue, Adrian Gheorghe, Janani Muhunthan, Michael Engelgau, Ajay Mahal, Ulla Griffiths, Diane McIntyre, Qingyue Meng, and et al. 2018. Action to Address the Household Economic Burden of Non-Communicable Diseases. *The Lancet* 391: 2047–58. [CrossRef]

Jha, Prabhat, Renu Joseph, David Li, Cindy Gauvreau, Ian Anderson, Patricia Moster, and Sekhar Bonu. 2012. *Tobacco Taxes: A Win-Win Measure for Fiscal Space and Health. November 2012.* Manila: Asian Development Bank.

Jha, Prabhat, Mary MacLennan, Frank J. Chaloupka, Ayda Yurekli, Chintanie Ramasundarahettige, Krishna Palipudi, Witold Zatonksi, Samira Asma, and Prakash C. Gupta. 2015. *Global Hazards of Tobacco and the Benefits of Smoking Cessation and Tobacco Taxes.* Washington, DC: World Bank.

John, Rijo M. 2008. Crowding out Effect of Tobacco Expenditure and Its Implications on Household Resource Allocation in India. *Social Science & Medicine* 66: 1356–67. [CrossRef]

John, Rijo M., Hai-Yen Sung, Wendy B. Max, and Hana Ross. 2011. Counting 15 Million More Poor in India, Thanks to Tobacco. *Tobacco Control* 20: 349. [CrossRef]

Karki, Yagya B., Kiran Dev Pant, and Badri Raj Pande. 2003. *A Study on the Economics of Tobacco in Nepal.* Washington, DC: World Bank.

Kopelman, P. 2007. Health Risks Associated with Overweight and Obesity. *Obesity Reviews* 8: 13–17. [CrossRef]

Lehnert, Thomas, Pawel Streltchenia, Alexander Konnopka, Steffi G. Riedel-Heller, and Hans-Helmut König. 2015. Health Burden and Costs of Obesity and Overweight in Germany: An Update. *The European Journal of Health Economics* 16: 957–67. [CrossRef]

Leigh, J. Paul, and Mark C. Berger. 1989. Effects of Smoking and Being Overweight on Current Earnings. *American Journal of Preventive Medicine* 5: 8–14. [CrossRef]

Levesque, Jean-Frédéric, Slim Haddad, Delampady Narayana, and Pierre Fournier. 2007. Affording What's Free and Paying for Choice: Comparing the Cost of Public and Private Hospitalizations in Urban Kerala. *The International Journal of Health Planning and Management* 22: 159–74. [CrossRef]

Levy, David T., Frank J. Chaloupka, and Joseph Gitchell. 2004. The Effects of Tobacco Control Policies on Smoking Rates: A Tobacco Control Scorecard. *Journal of Public Health Management and Practice* 10: 338–53. [CrossRef]

Liu, Yuanli, Keqin Rao, Teh-wei Hu, Qi Sun, and Zhenzhong Mao. 2006. Cigarette Smoking and Poverty in China. *Social Science & Medicine* 63: 2784–90. [CrossRef]

Low, Serena, Mien Chew Chin, and Mabel Deurenberg-Yap. 2009. Review on Epidemic of Obesity. *Annals Academy of Medicine Singapore* 38: 57.

Madore, Amy, Julie Rosenberg, and Rebecca Weintraub. 2015. Harvard Business Publishing "Sin Taxes" and Health Financing in the Philippines. Harvard Business Publishing. Available online: https://www.globalhealthdelivery.org/case-collection/case-studies/asia-and-middle-east/sin-taxes-and-health-financing-in-the-philippines (accessed on 4 August 2020).

Malik, Aaqib Habib, Yasir Akram, Suchith Shetty, Senada Senda Malik, and Valentine Yanchou Njike. 2014. Impact of Sugar-Sweetened Beverages on Blood Pressure. *The American Journal of Cardiology* 113: 1574–80. [CrossRef] [PubMed]

Malik, Vasanti S., Matthias B. Schulze, and Frank B. Hu. 2006. Intake of Sugar-Sweetened Beverages and Weight Gain: A Systematic Review. *The American Journal of Clinical Nutrition* 84: 274–88. [CrossRef] [PubMed]

McCambridge, Jim, Melissa Mialon, and Ben Hawkins. 2018. Alcohol Industry Involvement in Policymaking: A Systematic Review. *Addiction* 113: 1571–84. [CrossRef]

Misra, Anoop, and Lokesh Khurana. 2008. Obesity and the Metabolic Syndrome in Developing Countries. *The Journal of Clinical Endocrinology & Metabolism* 93 S 1: s9–s30.

Mourao, Denise M., Josefina Bressan, Wayne W. Campbell, and Richard David Mattes. 2007. Effects of Food Form on Appetite and Energy Intake in Lean and Obese Young Adults. *International Journal of Obesity* 31: 1688–95. [CrossRef]

Nargis, Nigar, Ummul H Ruthbah, AKM Ghulam Hussain, Geoffrey T Fong, Iftekharul Huq, and S. M. Ashiquzzaman. 2014. The Price Sensitivity of Cigarette Consumption in Bangladesh: Evidence from the International Tobacco Control (ITC) Bangladesh Wave 1 (2009) and Wave 2 (2010) Surveys. *Tobacco Control* 23 S1: i39–i47. [CrossRef]

NCI (U.S. National Cancer Institute), and WHO (World Health Organization). 2016. *The Economics of Tobacco and Tobacco Control. National Cancer Institute Tobacco Control Monograph 21*; NIH Publication No. 16-CA-8029A. Bethesda: U.S. Department of Health and Human Services, National Institutes of Health, National Cancer Institute, Geneva: World Health Organization.

Niessen, Louis W., Diwakar Mohan, Jonathan K. Akuoku, Andrew J. Mirelman, Sayem Ahmed, Tracey P. Koehlmoos, Antonio Trujillo, Jahangir Khan, and David H. Peters. 2018. Tackling Socioeconomic Inequalities and Non-Communicable Diseases in Low-Income and Middle-Income Countries under the Sustainable Development Agenda. *The Lancet* 391: 2036–46. [CrossRef]

Nugent, Rachel, Melanie Y Bertram, Stephen Jan, Louis W Niessen, Franco Sassi, Dean T Jamison, Eduardo González Pier, and Robert Beaglehole. 2018. Investing in Non-Communicable Disease Prevention and Management to Advance the Sustainable Development Goals. *The Lancet* 391: 2029–35. [CrossRef]

O'Donoghue, Ted, and Matthew Rabin. 2006. Optimal Sin Taxes. *Journal of Public Economics* 90: 1825–49. [CrossRef]

OECD. 2019. *The Heavy Burden of Obesity: The Economics of Prevention*. OECD Health Policy Studies. Paris: OECD Publishing. [CrossRef]

Onder, Zeynep. 2002. *The Economics of Tobacco in Turkey: New Evidence and Demand Estimates*. Washington, DC: World Bank.

Pampel, Fred C., Justin T. Denney, and Patrick M. Krueger. 2012. Obesity, SES, and Economic Development: A Test of the Reversal Hypothesis. *Social Science & Medicine* 74: 1073–81. [CrossRef]

Paraje, Guillermo, and Daniel Araya. 2018. Relationship between Smoking and Health and Education Spending in Chile. *Tobacco Control* 27: 560–67. [CrossRef]

Parry, Charles D., Jayadeep Patra, and Jürgen Rehm. 2011. Alcohol Consumption and Non-Communicable Diseases: Epidemiology and Policy Implications. *Addiction* 106: 1718–24. [CrossRef]

Pigou, Arthur Cecile. 1920. *The Economics of Welfare*. London: Macmillan.

Pitayatienanan, Paiboon, Rukmanee Butchon, Jomkwan Yothasamut, Wichai Aekplakorn, Yot Teerawattananon, Naeti Suksomboon, and Montarat Thavorncharoensap. 2014. Economic Costs of Obesity in Thailand: A Retrospective Cost-of-Illness Study. *BMC Health Services Research* 14: 146. [CrossRef]

Pu, Cheng-yun, Virginia Lan, Yiing-Jenq Chou, and Chung-fu Lan. 2008. The Crowding-out Effects of Tobacco and Alcohol Where Expenditure Shares Are Low: Analyzing Expenditure Data for Taiwan. *Social Science & Medicine* 66: 1979–89. [CrossRef]

Ranaweera, Sajeeva, Hemantha Amarasinghe, Nadeeka Chandraratne, Montarat Thavorncharoensap, Thushara Ranasinghe, Sumudu Karunaratna, Dinesh Kumara, Benjarin Santatiwongchai, Usa Chaikledkaew, Palitha Abeykoon, and et al. 2018. Economic Costs of Alcohol Use in Sri Lanka. *PLoS ONE* 13: e0198640. [CrossRef]

Ravishankar, A. K. 2012. Is India Shouldering a Double Burden of Malnutrition? *Journal of Health Management* 14: 313–28. [CrossRef]

Rehm, Jürgen, and I. Rossow. 2001. The Impact of Alcohol Consumption on Work and Education. In *Mapping the Social Consequences of Alcohol Consumption*. Edited by Harald Klingemann and Gerhard Gmel. Dordrecht: Springer, pp. 67–77. [CrossRef]

Renna, Francesco. 2008. Alcohol Abuse, Alcoholism, and Labor Market Outcomes: Looking for the Missing Link. *ILR Review* 62: 92–103. [CrossRef]

Sarntisart, Isra. 2003. *An Economic Analysis of Tobacco Control in Thailand*. Washington, DC: World Bank.

Sassi, Franco, Annalisa Belloni, Andrew J. Mirelman, Marc Suhrcke, Alastair Thomas, Nisreen Salti, Sukumar Vellakkal, Chonlathan Visaruthvong, Barry M Popkin, and Rachel Nugent. 2018. Equity Impacts of Price Policies to Promote Healthy Behaviours. *The Lancet* 391: 2059–70. [CrossRef]

Saxena, Akshar, Adam D. Koon, Leizel Lagrada-Rombaua, Imelda Angeles-Agdeppa, Benjamin Johns, and Mario Capanzana. 2019. Modelling the Impact of a Tax on Sweetened Beverages in the Philippines: An Extended Cost–Effectiveness Analysis. *Bulletin of the World Health Organization* 97: 97. [CrossRef] [PubMed]

Semba, Richard D., Leah M. Kalm, Saskia de Pee, Michelle O. Ricks, Mayang Sari, and Martin W Bloem. 2007. Paternal Smoking Is Associated with Increased Risk of Child Malnutrition among Poor Urban Families in Indonesia. *Public Health Nutrition* 10: 7–15. [CrossRef]

Siahpush, Mohammad, Melanie A. Wakefield, Matt J. Spittal, Sarah J. Durkin, and Michelle M. Scollo. 2009. Taxation Reduces Social Disparities in Adult Smoking Prevalence. *American Journal of Preventive Medicine* 36: 285–91. [CrossRef] [PubMed]

Smith, Katherine, and Jon Foster. 2014. *Alcohol, Health Inequalities and the Harm Paradox: Why Some Groups Face Greater Problems despite Consuming Less Alcohol.* London: Institute of Alcohol Studies.

Son, Chong Hwan, and Kudret Topyan. 2011. The Effect of Alcoholic Beverage Excise Tax on Alcohol-Attributable Injury Mortalities. *The European Journal of Health Economics* 12: 103–13. [CrossRef] [PubMed]

Sornpaisarn, Bundit, Kevin Shield, Joanna Cohen, Robert Schwartz, and Jürgen Rehm. 2013. Elasticity of Alcohol Consumption, Alcohol-Related Harms, and Drinking Initiation in Low-and Middle-Income Countries: A Systematic Review and Meta-Analysis. *The International Journal of Alcohol and Drug Research* 2: 45–58. [CrossRef]

Staff, Jeremy, Megan E. Patrick, Eric Loken, and Jennifer L. Maggs. 2008. Teenage Alcohol Use and Educational Attainment. *Journal of Studies on Alcohol and Drugs* 69: 848–58. [CrossRef]

Stenberg, Karin, Riku Elovainio, Dan Chisholm, Daniela Fuhr, Anne-Marie Perucic, Dag Rekve, and Ayda Yurekli. 2010. *Responding to the Challenge of Resource Mobilization-Mechanisms for Raising Additional Domestic Resources for Health.* World Health Report. Geneva: World Health Organization.

Summan, Amit, and Ramanan Laxminarayan. 2019. *Modeling the Impact of Tobacco, Alcohol, and Sugary Beverage Tax Increases on Health and Revenue.* Backgroun Paper for the Task Force on Fiscal Policy and Health. New York: Bloomberg Philantropies. Available online: https://www.bbhub.io/dotorg/sites/2/2019/04/Modeling-the-Impact-of-Tobacco-Alcohol-and-Sugary-Beverage-Tax-Increases-on-Health-and-Revenue.pdf (accessed on 8 November 2019).

Summers, Lawrence H. 2018. Taxes for Health: Evidence Clears the Air. *The Lancet* 391: 1974–76. [CrossRef]

Suwa, Kiyomi, Natalia M. Flores, Reiko Yoshikawa, Rei Goto, Jeffrey Vietri, and Ataru Igarashi. 2017. Examining the Association of Smoking with Work Productivity and Associated Costs in Japan. *Journal of Medical Economics* 20: 938–44. [CrossRef]

Thavorncharoensap, Montarat, Yot Teerawattananon, Jomkwan Yothasamut, Chanida Lertpitakpong, and Usa Chaikledkaew. 2009. The Economic Impact of Alcohol Consumption: A Systematic Review. *Substance Abuse Treatment, Prevention, and Policy* 4: 20. [CrossRef] [PubMed]

The Lancet. 2017. Addressing the Vulnerability of the Global Food System. *The Lancet* 390: 95. [CrossRef]

The Task Force on Fiscal Policy and Health. 2019. *Health Taxes to Save Lives: Employing Effective Excise Taxes on Tobacco, Alcohol, and Sugary Beverages.* New York: Bloomberg Philanthropies. Available online: https://www.bbhub.io/dotorg/sites/2/2019/04/Health-Taxes-to-Save-Lives-Report.pdf (accessed on 9 October 2019).

Thow, Anne Marie, Shauna M. Downs, Christopher Mayes, Helen Trevena, Temo Waqanivalu, and John Cawley. 2018. Fiscal Policy to Improve Diets and Prevent Noncommunicable Diseases: From Recommendations to Action. *Bulletin of the World Health Organization* 96: 201. [CrossRef]

Townsend, J., P. Roderick, and J. Cooper. 1994. Cigarette Smoking by Socioeconomic Group, Sex, and Age: Effects of Price, Income, and Health Publicity. *BMJ* 309: 923–27. [CrossRef] [PubMed]

Traversy, Gregory, and Jean-Philippe Chaput. 2015. Alcohol Consumption and Obesity: An Update. *Current Obesity Reports* 4: 122–30. [CrossRef]

Tsikomia, Amaïde Arsan Miriarison, and Daniela Şarpe. 2012. Alcohol Consumption: Measuring the Risk of Household Poverty (Case of the Urban District of Toamasina-Madagascar). *Economics and Applied Informatics* 2: 139–47.

UN (United Nations). 2015a. *Report of the Third International Conference on Financing for Development, Addis Ababa, 13–16 July 2015.* New York: United Nations.

UN (United Nations). 2015b. Transforming Our World: The 2030 Agenda for Sustainable Development. Working Papers. eSocialSciences. Available online: https://EconPapers.repec.org/RePEc:ess:wpaper:id:7559 (accessed on 18 October 2019).

USDHHS (U.S. Department of Health and Human Services). 2014. *The Health Consequences of Smoking—50 Years of Progress. A Report of the Surgeon General*; Atlanta: US Department of Health and Human Services, Centers for Disease Control and Prevention, National Center for Chronic Disease Prevention and Health Promotion, Office on Smoking and Health.

Van Walbeek, C. 2005. Tobacco Control in South Africa. *Promotion & Education* 4: 25–28.

Verguet, Stéphane, Cindy L. Gauvreau, Sujata Mishra, Mary MacLennan, Shane M. Murphy, Elizabeth D. Brouwer, Rachel A. Nugent, Kun Zhao, Prabhat Jha, and Dean T. Jamison. 2015. The Consequences of Tobacco Tax on Household Health and Finances in Rich and Poor Smokers in China: An Extended Cost-Effectiveness Analysis. *The Lancet Global Health* 3: e206–e216. [CrossRef]

Virk-Baker, Mandeep, Muhammad Jami Husain, and Mark Parascandola. 2019. Comparative Analysis of Diet and Tobacco Use among Households in Bangladesh. *Tobacco Prevention & Cessation* 5. [CrossRef]

Wagenaar, Alexander C., Amy L. Tobler, and Kelli A. Komro. 2010. Effects of Alcohol Tax and Price Policies on Morbidity and Mortality: A Systematic Review. *American Journal of Public Health* 100: 2270–78. [CrossRef]

Wagstaff, Adam. 2002. Poverty and Health Sector Inequalities. *Bulletin of the World Health Organization* 80: 97–105. [PubMed]

Wang, Y. Claire, John Pamplin, Michael W. Long, Zachary J. Ward, Steven L. Gortmaker, and Tatiana Andreyeva. 2015. Severe Obesity In Adults Cost State Medicaid Programs Nearly $8 Billion In 2013. *Health Affairs* 34: 1923–31. [CrossRef] [PubMed]

WCRF (World Cancer Research Fund International). 2018. *Building Momentum: Lessons on Implementing a Robust Sugar Sweetened Beverage Tax.* London: World Cancer Research Fund International. Available online: https://www.wcrf.org/sites/default/files/PPA-Building-Momentum-Report-2-WEB.pdf (accessed on 9 November 2019).

Weng, S. F., S. Ali, and J. Leonardi-Bee. 2013. Smoking and Absence from Work: Systematic Review and Meta-Analysis of Occupational Studies. *Addiction* 108: 307–19. [CrossRef] [PubMed]

WHO (World Health Organization). 2014. *Global Status Report on Alcohol and Health 2014.* Geneva: World Health Organization. Available online: https://www.who.int/substance_abuse/publications/global_alcohol_report/msb_gsr_2014_1.pdf?ua=1 (accessed on 8 November 2019).

WHO (World Health Organization). 2017. *World Health Statistics 2017: Monitoring Health for the SDGs.* Geneva: World Health Organization. Available online: https://www.who.int/gho/publications/world_health_statistics/2017/en/ (accessed on 8 August 2019).

WHO (World Health Organization). 2018a. *Global Health Estimates 2016: Disease Burden by Cause, Age, Sex, by Country and by Region, 2000–2016. Geneva: World Health Organization.* Geneva: World Health Organization. Available online: https://www.who.int/healthinfo/global_burden_disease/estimates/en/index1.html (accessed on 8 November 2019).

WHO (World Health Organization). 2018b. *Global Status Report on Alcohol and Health 2018.* Geneva: World Health Organization. Available online: https://apps.who.int/iris/bitstream/handle/10665/274603/9789241565639-eng.pdf?ua=1 (accessed on 8 November 2019).

WHO (World Health Organization). 2019a. *Global Health Observatory (GHO) Data.* Geneva: World Health Organization. Available online: https://www.who.int/gho/ncd/risk_factors/overweight/en/ (accessed on 25 October 2019).

WHO (World Health Organization). 2019b. *Global Information System on Alcohol and Health (GISAH).* Geneva: World Health Organization. Available online: http://apps.who.int/gho/data/node.gisah.GISAHhome?showonly=GISAH (accessed on 8 November 2019).

WHO (World Health Organization). 2019c. *WHO Report on the Global Tobacco Epidemic, 2019: Offer Help to Quit Tobacco Use.* Geneva: World Health Organizaton.

Transforming Regional Agrifood Productions to Challenge NCDs—From the DiMeSa Study to the PASSI Project and Beyond

Giuseppe Carruba

1. Foreword: Can Two Emergencies Make an Opportunity?

In our country, like in many other regions worldwide, two major, apparently unrelated, crises are deeply impacting the wellbeing and health of the population and the economy. There is, however, substantial indication that these two distinct emergencies are strictly interdependent, if not faces of the same coin, and that, thereof, systemic, cross-sectional strategies, implemented for and merging critical issues of both sides, could be highly effective in providing an answer, in the medium- and long-term, to this highly demanding, global occurrence.

The Italian agrifood system is of considerable importance both in terms of turnover and number of businesses and employment for the characterization of the Made in Italy and, more in general, the Italian lifestyle across the world. The Italian food industry faces today a phase of profound evolution based, on one hand, on the internal dynamics of the system and, on the other, on the more general process of globalization that profoundly affects the entire world economy. These changes are influenced by and, in turn, influence new consumer behavior trends, though factors associated with regional level and individual cultural diversities significantly and systematically impact impulsive purchasing behavior (Kacen and Lee 2002).

The radical changes in the production, commercial and distribution systems related to food products have generated socio-economic changes that have in turn modified consumers' eating habits (Greendex 2014). In recent decades, we have witnessed weighty social phenomena, such as the destructuring of the family nucleus, the increasing inclusion of women in both the labor market and economic activities, the spread of the market nonstop opening hours and the steady increase in out-of-home food consumption, which have all determined a profound reorganization of life rhythms and contributed to generating significant changes in eating patterns. The latter are also featured by an ever increasing degree of processing, hence favoring a decrease in the time allocated for food preparation and consumption and determining a consequential preference for products with a higher content of additional services that consumers are willing to purchase at higher prices.

Another phenomenon that has expanded in recent years is the demand for safe food products with stable quality over time, providing the consumer with a high level of satisfaction based on taste (Grunert 2005). This is, in turn, closely connected to other important aspects, such as the impact on both individual and collective health status and well-being, the genuineness and naturalness of the agricultural raw materials used, the link with the territory, its history and traditions, factors that have all contributed to the realization of systems for the traceability of food products and the valorization of different food and wine cultures dispersed throughout the national and regional territories, including our own. Furthermore, recent changes in the Italian agrifood system originate from the transformation of the distribution systems, which have become increasingly similar to those already established in most advanced European countries, whereby, along with the considerable gigantism of major companies in the large-scale retail trade that compete on prices and dictate the conditions for other companies to enter the market, there are small/medium-sized agricultural and agrifood enterprises, generally characterized by low bargaining power and limited capacity to occupy significant market segments. Moreover, these companies, being generally based on atomistic structures, suffer the ensuing difficulty of developing innovative, competitive products on the market at large.

As far as our regional agrifood sector is concerned, although it has the highest number of enterprises in the country, with the greatest percentages of agricultural area used for vineyards, fruits and vegetables cultures, it has, on average, one of the lowest company dimensions, in terms of either hectares or employees per company (ISTAT 2010). Furthermore, regional investments in the agrifood sector lie well below the national average (nearly 50%), with the agriculture added value in Sicily plateauing and occupation units in the agroindustry having decreased over the last decade (INEA 2013). Additionally, small/medium enterprises (SMEs) in the field are often featured by a limited innovation potential, a poor integration with public-private research institutions and an insufficient systematization and organization of the existing resources in an extended territorial networking. This results in increasing difficulties for SMEs to run both domestic and foreign markets with characteristics of quality and competitiveness.

According to WHO, noncommunicable diseases (NCDs), mostly represented by cardiovascular disease, cancer, chronic respiratory disease and diabetes, accounted for 71% of the 57 million deaths worldwide in 2016 (WHO 2018a). In our own region, NCDs account for nearly 80% of all causes of death in both sexes (ReNCaM 2014).

However, the global risk of dying from any one of these four major NCDs in people aged 30 to 70 years has steadily decreased in the last two decades (from 22% in 2000 to 18% in 2016) (WHO 2018b). On the other hand, several epidemiological studies clearly indicate that all Western countries, including Italy, are witnessing real epidemics of these NCDs and both adult and childhood obesity, mostly because of the dramatic changes that have occurred in both food systems and consumer eating habits and behavior globally (Hunter and Reddy 2013). Furthermore, the appearance of many chronic diseases is occurring today at an average age earlier than ever, with a progressive leftward shift in the onset age of various illnesses, including diabetes, cancer and obesity (Gale 2002).

Paradoxically, or subsequently if you like, these changes have produced the most profound and rapid consequences in Southern European countries, including Italy, Greece, Portugal and Spain, eventually leading to a significant increase in incidence rates of NCDs, along with high percentages of overweight and/or obese children (Wijnhoven et al. 2014), as a result of cultural, social, economic and genetic transformations. The progressive decline in mortality from NCDs, combined with the rise in their incidence worldwide, has produced an alarming "scissor" phenomenon, consisting in a remarkable and continuous increase in the total number of chronically ill individuals, that is to say of NCD prevalence (Hunter and Reddy 2013). This trend causes economic, social and health issues of outstanding relevance worldwide and eventually led the World Health Organization (WHO), along with all major health institutions, to launch a pluriannual action plan (2013–2020) for the primary prevention and control of NCDs based on intersectorial strategies targeting the removal of key risk factors for these diseases, including tobacco use, air pollution, physical inactivity, harmful use of alcohol and, notably, unhealthy diet (WHO 2013).

In this framework, the increasingly high prevalence of NCDs is basically a consequence of the transition from traditional to contemporary dietary patterns and physical activity. This shift, often named nutritional transition, is based on the increasingly large availability of inexpensive, high calorie-dense food, quite often at the expense of biodiverse, local and healthier products. In particular, processed and ultra-processed food, rich in salt, refined sugar or sweeteners, saturated and trans fats, is largely outpacing healthy food, rich in micronutrients, including fresh fruits, vegetables, legumes and nuts. Furthermore, while the intake of unrefined, whole grains is fading and that of fruits and vegetables remains largely insufficient, the consumption of meat, dairy products and sweetened drinks has markedly expanded in most regions around the world. In a recent Italian study, Giampaoli and colleagues compared the food consumption of a self-reported survey recorded in

1960 in a southern Italian village (Nicotera) with that of an OEC/HES survey carried out in 2008–2012 on 1968 men and 2062 women, aged 40–59 years. The authors found that the consumption of cereals has dropped to less than 40% in both sexes, while the intake of meat, dairy products, milk and sweets has increased from 2- up to 4-fold (Giampaoli et al. 2015).

By definition, a food system comprises "all the elements (environment, people, inputs, processes, infrastructure, institutions) and activities that relate to the pre-production, production, processing, distribution, preparation and consumption of food and the outputs of these activities, including socioeconomic and environmental outcomes" (High Level Panel of Experts on Food Security and Nutrition 2017). In this wider context, individuals procure and consume food in the food environment, defined as "the interface that encompasses external dimensions such as the availability, prices, vendor and product properties, and promotional information, as well as personal dimensions such as the accessibility, affordability, convenience and desirability of food sources and products" (Turner et al. 2018).

A number of disparate factors determine the structure of food systems (see Figure 1).

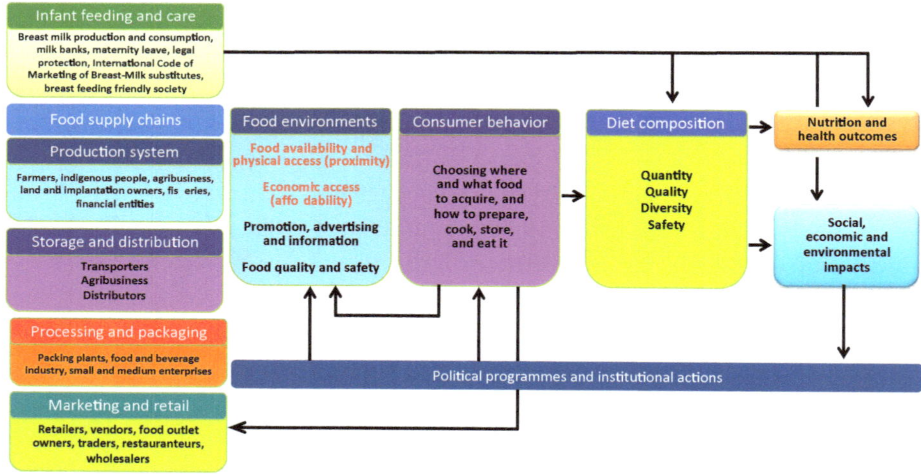

Figure 1. Schematic representation of food system and its outcomes. Source: Modified from Branca et al. (2019), used with permission.

They relate to a vast array of drivers and policies, as described by Branca and colleagues (Branca et al. 2019). The former include demographic issues, globalization, urbanization and climate changes. The latter comprise environmental, agricultural, commercial and economic policies, either locally or globally. According to Turner and associates (Turner et al. 2018), food environments consist of both external and personal domains, whereby a variety of dimensions coexist and shape the environment: the external domain comprises subjects relevant to food availability and prices, characteristics of vendors and products, marketing and its regulation; the personal domain is based on individual dimensions, including food accessibility, affordability, convenience and desirability. The authors consider that people's food acquirement and consumption is generated by a series of intricate interactions between these domains and their assorted dimensions.

In the food environment, consumer behavior has a central role and is subjected to a multiplicity of influencing factors, belonging to cultural background and individual knowledge; personal preferences; food availability and accessibility, either physical (reachability) or economic (affordability); time and effort allocable for food purchase; preparation; cooking; and consumption. In recent decades, consumer behavior has been the primary target of aggressive marketing strategies of major food companies and large retail chains, eventually corrupting the food system and delivering mostly inexpensive, unhealthy food, with devastating effects on public health, especially for low-income, developing countries and vulnerable and marginalized populations. Today, although citizens are are literally soaked in a liquid milieu of innumerable news, information, reports, forums, blogs, advertisements and mass and social media, all available on the internet, quite often most of them come from an extremely large and doubtful variety of sources that can heavily affect and/or undermine consumer behavior and beliefs. Luckily enough, however, the basic idea that food is important for human health has pervaded the general population at large and generated an increasingly greater demand for healthy food, with a greater intake of fruits and vegetables and a growing interest in products containing superfoods as healthful ingredients. In this highly fragmented and complicated context, there is an urgent need to develop policies, programs and campaigns for the nutritional counseling and education of citizens based on scientific, solid evidence and shared, verifiable information.

Based on this combined consideration, promoting both the production and competitiveness of traditional food products—in regional, domestic and international markets, through a series of activities aimed at increasing their health and/or nutraceutical potential, to clinically validate their effects on both health and chronic disease(s); to enable the rapid technological transfer and industrial development of either processes or products; and to exploit large, population-based interventions of nutritional and behavioral education to develop individual knowledge and competences (empowerment) concerning lifestyle, nutrition and health—would represent a systemic strategy of high impact in the short, medium and long term for important expected outcomes from an economic, technological and healthcare standpoint.

2. The DiMeSa Study

The DiMeSa study[1], where DiMeSa stands for **Dieta Mediterranea e Salute**, simply Mediterranean Diet and Health, was funded by the Italian Ministry of University and Research (MIUR) to the leader Institution, the AgroBioPesca Technology Cluster. The project, which ran from October 2012 up to December 2015, was named "DiMeSa—Valorization of typical products of Mediterranean Diet and their use for health and nutraceutical purposes" and aimed at increasing the attractiveness and competitiveness, in either the domestic or international market, of traditional products from major regional agrifood chains. In particular, the main objective of the DiMeSa project was, using different approaches, to develop and exploit innovative industrial research and experimental activities that would eventually lead to improving the health potential of traditional food products and that, at the same time, would scientifically validate the existing relationship between these products or their components and health, both in terms of maintaining, both individually and collectively, a wellbeing condition and, especially, of the primary prevention of NCDs.

As illustrated in Figure 2, the DiMeSa project was arranged into 4 major objectives, precisely:

1. the analysis and identification of traditional food processes and the development of innovative biotechnological protocols for the production of food with high nutritional and health potential, including extra virgin olive oil, cereals or vegetables and their transformed derivatives;

2. the definition and implementation of procedures and methodological approaches for the production of functional foods (extra virgin olive oil, pasta, juices) through their combination (functionalization) with natural substances and/or plant/byproducts extracts with high health potential and their distribution through innovative vending machines;

3. the clinical validation of specific health claims through the implementation of randomized, controlled clinical trials to assess the health effects of selected

---

[1]   Based also on previously published work by Carruba et al. (2016).

functional food products on cohorts of either healthy, high-risk or diseased study-subjects through the evaluation of the impact of dietary intervention on some clinical and biomolecular end-points, such as: (a) anthropometric measures; (b) immunological markers of inflammation; (c) oxidative stress and endothelial function; (d) hormonal profiles and gene/miRNA expression;

4. the economic evaluation of the concept, traceability and industrial scale-up of either prototypal products or processes aiming to allow their immediate industrialization and successful marketing.

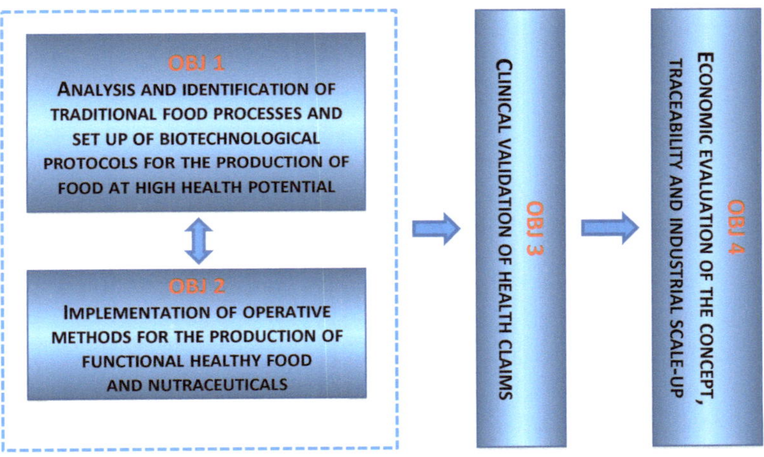

Figure 2. Framework of the Dieta Mediterranea e Salute (DiMeSa) project and its 4 major objectives. Source: Figure by the author.

The DiMeSa project was a multicenter study, characterized by a large regional partnership, which included universities and other public research institutions, on one hand, and small/medium enterprises (SMEs) in the field, on the other (see Figure 3).

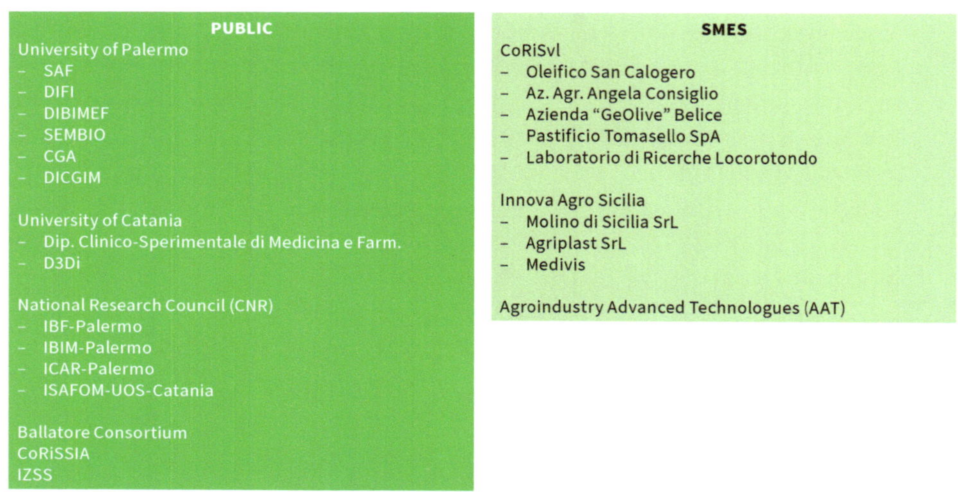

Figure 3. Partnership of the DiMeSa project, including public research bodies and small/medium enterprises (SMEs). Source: Figure by the author.

In the framework of the DiMeSa project, we conducted two main clinical trials, both belonging to Objective 3 of the project.

In the first randomized study, we assessed the health impact of monocultivar extra virgin olive oils (EVOs) on 2 cohorts of study subjects represented by healthy postmenopausal women and patients with breast cancer. All study subjects were recruited at the Azienda di Rilievo Nazionale e di Alta Specializzazione (ARNAS)—Civico, Di Cristina, Benfratelli (CDB). Overall, 103 healthy postmenopausal women and 35 breast cancer patients were enrolled in the study. Two different mono-cultivar EVOs, one at a lower (*Biancolilla* cultivar, BL) and one at a higher (*Cerasuola* cultivar, CS) content of polyphenols and oleocanthal, were used in the study; both EVOs were produced by the Department of Agriculture and Forestry Sciences (SAF) of Palermo University, under the supervision of Prof. Tiziano Caruso. As regards healthy women, after an initial one-week wash-out period ("no EVO" week), the subjects consumed a daily amount of 30 mL of the BL EVO for 4 weeks, followed by another one week wash-out period ("no EVO") and an additional 4 weeks intervention with the CS EVO, as illustrated in Figure 4. Conversely, breast cancer patients were randomized into one BL EVO and one CS EVO intervention group that consumed daily amounts of 30 mL of either BL or CS EVO for 4 weeks (see Figure 4). Both healthy and breast cancer study subjects, before and after any EVO intervention, undertook the following: (a) compiled a food frequency questionnaire originally developed for the EPIC study (Pisani et al. 1997);

(b) measured anthropometric indexes, including height, weight and waist-to-hip ratio; (c) were administered psychometric tests (HADS, SF-36); (d) collected both fasting blood samples and 12 h urine samples. The latter samples were used to determine the potential effect of dietary intervention on an array of both plasmatic and serum biomarkers, the expression profiles of a set of previously selected genes, the whole miRNome and the urinary profile of sex steroid hormones.

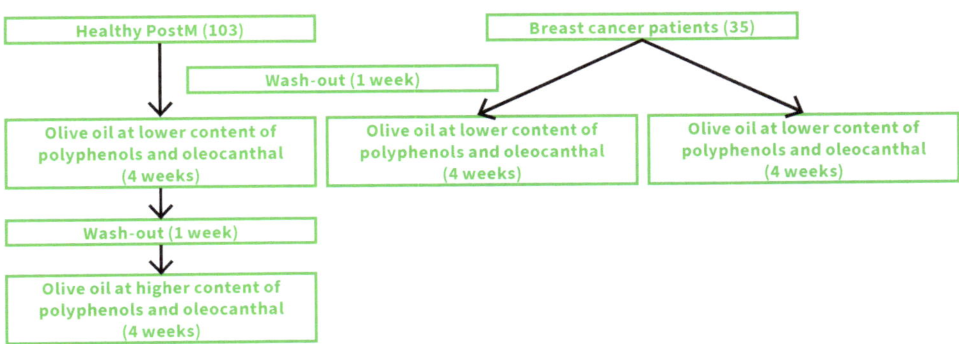

Figure 4. Flow chart of a randomized clinical trial in the DiMeSa project to assess the effects of selected monocultivar extra virgin olive oil on selected parameters in both healthy postmenopausal women and breast cancer patients. For explanation, see text. Source: Figure by the author.

Biological samples (plasma, serum and urine) were collected and stored in a biobank until analyses. A web-based study database was also created and data processed using advanced statistical analysis and appropriate software.

As reported in Table 1, the consumption of BL EVO resulted in significant changes of various plasmatic biomarkers in both healthy subjects and breast cancer patients. In particular, the reduction in glycemia, insulinemia and total cholesterol levels appears to be of special interest.

On the other hand, the consumption of CS EVO produced several modifications in selected biomarkers (see Table 1), including a significant increase in HDL cholesterol and reduction in LDL cholesterol. Interestingly, this EVO also induced a marked decrease in the plasmatic levels of estradiol. It appears noteworthy that, when comparing the two EVOs, BL EVO appeared to be more effective in reducing glycemia, while CS EVO proved to be more effective in decreasing plasmatic estradiol (see Table 2).

Table 1. Effects of BL and CS extra virgin olive oil (EVO) on plasmatic biomarkers in both healthy postmenopausal women and breast cancer patients.

BL Extra Virgin Olive Oil				CS Extra Virgin Olive Oil			
Variable	Baseline	After	*p*-Value *	Variable	Baseline	After	*p*-Value *
Azotemia	30.85	28.48	0.002	Cretininemia	0.69	0.63	<0.001
Uricemia	4.17	4.29	0.001	Uricemia	4.23	4.43	0.002
Glycemia	85.35	83.59	0.021	Glycemia	89.16	88.48	0.023
Insulinemia	10.33	8.79	<0.001	Glycated hemoglobin	5.64	5.51	<0.001
Total cholesterol	207.48	197.12	<0.001	HDL cholesterol	57.87	59.31	0.023
Gamma GT	21.90	24.50	0.001	LDL cholesterol	119.62	102.12	0.047
Total Proteinemia	7.00	6.91	0.005	Testosterone	0.39	0.36	0.033
Sideremia	76.95	67.02	<0.001	Estradiol	31.40	23.95	0.002

* paired T test, ANOVA. BL, *Biancolilla* cultivar; CS, Cerasuola cultivar. Source: Carruba et al. 2016.

Table 2. Comparison of the effects of BL and CS EVO on glycemia and estradiol levels in both healthy postmenopausal women and breast cancer patients.

Variable	BL EVO	CS EVO	*p*-Value
Glycemia	83.59	88.48	0.023
Estradiol	37.21	23.95	0.027

BL, *Biancolilla* cultivar; CS, Cerasuola cultivar. Source: Carruba et al. 2016.

This would imply that different EVOs may have distinct impacts on either glycemic control or hormonal (sex steroid) status, also depending on the cultivar and on the phenology of fruit ripening (the earlier the stage, the greater the content of polyphenols).

The results of gene expression, microRNA profiling and patterns of urinary sex steroids are currently under complex analysis and are awaited with great expectation and interest.

In the context of the DiMeSa project, we conducted another clinical trial to assess the impact of a pasta product supplemented with an extract of cladodes of Opuntia Ficus Indica (OFI) on the lipid profiles of subjects presenting with at least one component of the metabolic syndrome (MS).

As illustrated in Figure 5, eighty-six healthy subjects of both sexes, among over 2500 employees, participated voluntarily in the study at the Division of Research and Internationalization of the ARNAS-Civico, Di Cristina, Benfratelli healthcare comprehensive center (Palermo, Italy) for 8 months.

Out of these 86 subjects, 52 aged 40–65 years (13 male and 36 female; age: 56 ± 5 years) were recruited in the study based on the inclusion criterion of presenting with one or more risk factors of MS according to the American heart Association (AMA), precisely: (a) fasting glucose ≥ 100 mg/dL (or receiving drug therapy for hyperglycemia); (b) blood pressure ≥ 130/85 mm Hg (or receiving drug therapy for hypertension); (c) triglycerides ≥ 150 mg/dL (or receiving drug therapy for hypertriglyceridemia); (d) HDL-C < 40 mg/dL in men or <50 mg/dL in women (or receiving drug therapy for reduced HDL-C); (e) waist circumference ≥ 102 cm in men or ≥88 cm in women.

Dietary intervention consisted in the weekly consumption of 500 g of the dried pasta functionalized with a 3% soluble extract of OFI cladodes for a total of 4 weeks and was maintained as an add-on to the cardio–metabolic therapies already in use. At baseline and at the end of dietary intervention, all study subjects underwent the following: (1) anthropometric measures; (2) psychometric tests; (3) medical examination; (4) biochemical assessment of circulating biomarkers, with special emphasis on the LDL-C subfractions (conducted at the Unit of Diabetes and Cardiovascular Prevention at the University of Palermo, Italy—Proff. G. Montalto and M. Rizzo).

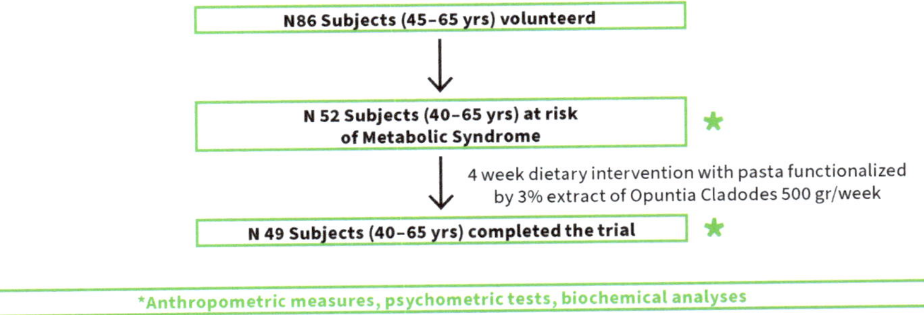

Figure 5. Flow chart of a randomized clinical trial in the DiMeSa project to assess the effects of pasta supplemented with a 3% extract of cladodes of Opuntia Ficus Indica on selected parameters in subjects at risk of metabolic syndrome. For explanation, see text. Source: Figure by the author.

In particular, the LipoPrint System (Quantimetrix Corporation, Redondo Beach, CA, USA) was used to separate and measure LDL-C subclasses, as described elsewhere

(Hoefner et al. 2001). LDL subclasses were divided into seven bands (LDL-1 to LDL-7, respectively), LDL-1 and -2 being defined as large LDL, while LDL-3 to -7 were defined as small LDL (Rizzo and Berneis 2006). At baseline, all participants were asked not to vary their food and/or physical activity during the period of the study. After a complete description of the study, written informed consent was obtained from all participants. The procedures adopted were approved by the Ethics Committees (EC) of both ARNAS-Civico (EC2) and the Policlinico University of Palermo (EC1). Overall, 49 subjects completed the study.

After 4 weeks dietary intervention with the OFI-supplemented pasta, a limited but significant decrease in waist circumference was observed, while neither body weight nor body mass index (BMI) showed any significant change (Table 3).

Table 3. Changes of biometric parameters after 4 weeks dietary intervention with Opuntia Ficus Indica (OFI)-supplemented pasta.

Variable	Baseline	4 Weeks	p Value *
Weight (kg)	69.5 (14.6) 67.8 (47.9–125.4)	69.8 (14.1) 69.1 (47.8–123.4)	0.8665
Waist circumference (cm)	92.3 (12.3) 92 (73–132)	91.4 (10.5) 91 (72–129)	0.0297
BMI (kg/m^2)	25.9 (3.9) 25.3 (20.8–41.4)	26.1 (3.7) 25.5 (20.8–40.8)	0. 8788

Values are mean (±SD) and median (min–max). * Wilcoxon paired test. Source: Carruba et al. 2016.

Furthermore, dietary intervention resulted in a statistically significant modification of several biochemical plasmatic parameters, including a significant reduction in blood urea nitrogen (BUN), creatinine, glucose, triglycerides, aspartate transaminase (AST) and sideremia; conversely, some hormonal parameters, including estrogens (estradiol and estrone) and sex hormone binding globulin (SHBG), showed a significant increase. As far as plasma lipids are concerned, total cholesterol (TC), low-density lipoprotein cholesterol (LDL-C) and high-density lipoprotein cholesterol (HDL-C) remained unchanged (see Table 4).

Interestingly, dietary intervention with the OFI-enriched pasta produced a significant increase in LDL-1 (from 49.6 ± 0.3 to 65.1 ± 0.2%, p = 0.0002) and a concomitant reduction in LDL-2 (40.1 ± 0.3 to 29.7 ± 0.2) and LDL-3 (8.3 ± 0.2 to 4.6 ± 0.1%, p = 0.0004). In addition, LDL-4 and LDL-5 also subclasses decreased, though differences were not statistically significant (see Table 5). No correlation was found between changes in sdLDL and any other metabolic parameter.

Table 4. Changes of plasmatic parameters after 4 weeks dietary intervention with OFI-supplemented pasta.

Variable	Baseline	4 Weeks	p Value
BUN (mg/dL)	32.7 (7.9) 33 (17–51)	43.3 (9.3) 43 (23–68)	<0.0001 *
Creatinine (mg/dL)	0.74 (0.1) 0.71 (0.52–1.33)	0.72 (0.1) 0.71 (0.43–1.25)	0.0244 *
Glycemia (mg/dL)	84.6 (12) 84 (58–128)	74.4 (14.2) 72 (60–158)	<0.0001 *
HbA1c (%)	5.4 (0.4) 5.3 (4.7–7.6)	5.4 (0.4) 5.3 (4.7–7.5)	0.9516 **
Total Cholesterol (mg/dL)	208.3 (36.1) 210 (115–267)	209.9 (35.6) 208 (129–266)	0.9620 **
HDL (mg/dL)	60.7 (13.9) 61 (30–88)	60.6 (13.3) 61 (30–88)	0.6672 **
LDL (mg/dL)	139.3 (32.9) 138 (65–195)	139.3 (33.6) 139 (78–200)	0.5274 **
Triglycerides (mg/dL)	104.6 (53.6) 82 (36–271)	92.8 (54.7) 71 (44–300)	0.0137 *
AST (mU/mL)	28.3 (8.2) 26 (17–63)	23.7 (8.4) 21 (15–62)	<0.0001 **
ALT (mU/mL)	22.2 (11.4) 18 (8–65)	22.6 (13.0) 19 (8–69)	0.8830 **
Gamma GT (U/L)	31.5 (27.5) 27 (11–167)	33.6 (39.2) 25 (11–268)	0.3855 **
Sideremia (µg/dL)	87.3 (21.3) 87 (45–134)	78.8 (28) 78 (26–158)	0.0255 *
Estradiol (pg/mL)	34.1 (27.2) 22 (7–102)	56.4 (58) 43 (25–432)	0.0055 **
SHBG (nmol/L)	66.5 (30.8) 60 (18–143)	81 (44) 66 (21–195)	<0.0001 **
Estrone (pg/mL)	23.7 (23.2) 15 (4–123)	77.7 (35.2) 64 (45–250)	<0.0001 **

Values are mean (±SD) and median (min–max). * Student's paired t test; ** Wilcoxon paired test. Source: Carruba et al. 2016.

Table 5. Effects of OFI-supplemented pasta on LDL-C subfractions.

LDL-C Fraction	Baseline	4 Weeks	p Value
LDL-1	49.6 ± 0.3	65.1 ± 0.2	0.0002
LDL-2	40.1 ± 0.3	29.7 ± 0.2	<0.0001
LDL-3	8.3 ± 0.2	4.6 ± 0.1	0.0004
LDL-4	1.3 ± 0.1	0.6 ± 0.0	0.2987
LDL-5	0.7 ± 0.1	0.0 ± 0.0	0.3223

Data represent average ± SD of percent values. Source: Carruba et al. 2016.

3. The PASSI Project

As a natural development of the DiMeSa project, we recently designed a new project called the Produzioni Agroalimentari Sostenibili Salutistiche e Identificabili in Sicilia—Sustainable, Healthy and Identifiable Agrifood Productions in Sicily (123 PASSI). The general framework of this project is illustrated in Figure 6.

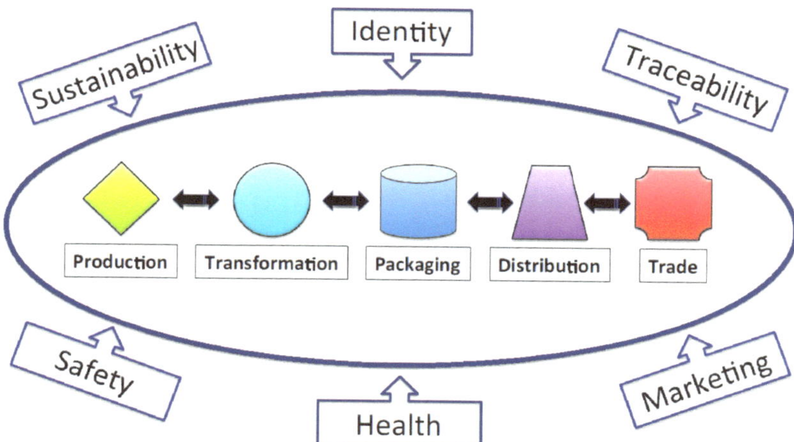

Figure 6. The conceptual framework of PASSI project. This picture portrays the overall goal of the project and its domains, from farm to market, that will be primary focus of the project's activities across major food chains. Additionally, the potential impact of key issues (from sustainability to marketing) on food system is also here highlighted as addressed in the project.

The general goal of the PASSI project consists not only of the systematization of experimental evidence and results accomplished with the DiMeSa project, both in terms of processes and products, but also to overcome its intrinsic limits by creating a collection of functional food products, traditionally belonging to the regional agrifood system, through research and development actions set up in different domains (production, processing, packaging, distribution and trade) throughout each supply chain (horizontal objectives)—along with those relevant to sustainability (environmental and industrial), traceability, food security, health potential and equity of access (vertical objectives)—generating knowledge of the supply chain and its identity value and, therefore, increasing consumer skills and decision-making capacities (empowerment) also through the use of algorithms (Blockchain, Distributed Ledger Technology) for the traceability of food supply chains and their health value (see Figure 7). The general objective of this project consists of the promotion and valorization in either domestic or international markets of high-quality functional food products typical of the traditional region, which are sustainable, traceable and safe, and whose health effect is proven in controlled and randomized clinical trials. The specific objectives comprise the development and exploitation of supply chain programs (from field to table; from farm to fork) that allow the production, processing, packaging, distribution and marketing of high-quality foods with a high health potential in major regional agrifood chains, precisely: (a) the extra virgin olive oil (EVO) chain; (b) cereal chain; (c) citrus chain; (d) fresh and dried fruit; (d) vine and wine chain. The activities to be developed consist of industrial research and experimental development activities, structured according to an immediate transfer of technologies and competences to the industrial setting for the production and realization/marketing of processes/products widely used and consumed in the Sicilian agro-industry. The project is divided into 8 implementation objectives (ORs), of which the first 5 (OR1–5) relate to food chain programs, while the remaining 3 (OR 6–8) to transversal research activities (see Figure 7).

The PASSI project is now being developed through its 9 implementation objectives and also integrated with additional, though highly relevant, aspects, with special emphasis on the use of agricultural byproducts and co-products to functionalize food of high consumption in the general population.

4. Conclusions and Perspectives

Doubtlessly, the primary prevention of NCDs through dietary interventions, in the wider context of health promotion through a life-course approach, as endorsed by WHO, requires the design and exploitation of systemic strategies, implemented both locally and globally, that would include radical changes in food systems and environments in order to provide identifiable, safe, healthy and affordable food to the current population. Virtually all aspects of food chains, including pre-production, productive processes, transformation, packaging, distribution, marketing and retail, should be renovated according to comprehensive approaches and policies shared at regional, national and international levels.

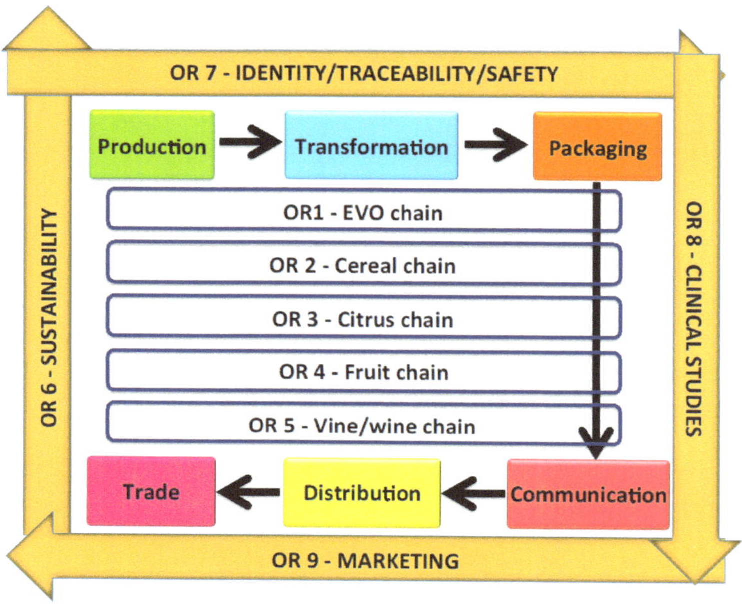

Figure 7. Structure of the 123 PASSI project. The project is divided into "horizontal" objectives (the traditional Mediterranean food chains: OR 1 to 5) and "vertical" objectives (from sustainability to marketing: OR 6 to 9), bridged by a sequential succession of activities (from production to trade) featuring the whole system. OR: implementation objective. Source: Figure by the author.

In particular, the adoption of incentives for primary producers and suppliers to produce and distribute healthy food sustainably should be accomplished. Conversely, fiscal policies to tax unhealthy, processed or ultra-processed food should also be implemented. Changes in food storage, transportation and distribution, aimed at providing easy access to nutritionally valuable, perishable food, should be introduced.

It is well recognized that health determinants consist of a vast array of personal, relational, social, economic and environmental factors that influence, individually and collectively, health status. According to Dahlgren and Whitehead (1991), socio-economic, cultural and environmental conditions include education, agriculture and food production as important pillars of wellbeing. In other words, health (in its in broadest sense) cannot be considered as an issue belonging solely to the welfare sector; conversely, it must be dealt with using intersectoral, multidimensional approaches and systemic strategies. In this paper, the PASSI project proposes a cross-sectoral model for the production and marketing of high health potential food products, addressing either internal (production, processing, packaging, distribution and trading) or external (sustainability, identity, traceability, safety, functionality and marketing) factors of major food chains. This approach, intrinsically, impacts various SDGs included in the 2030 Agenda for Sustainable Development, precisely: (1) Goal 2: end hunger, achieve food security and improved nutrition and promote sustainable agriculture; (2) Goal 3: ensure healthy lives and promote wellbeing for all at all ages; (3) Goal 12: ensure sustainable consumption and production patterns. Furthermore, it is indirectly related to the following SDGs: (4) Goal 4: ensure inclusive and equitable quality education and promote lifelong learning opportunities for all; (5) Goal 8: promote sustained, inclusive and sustainable economic growth, full and productive employment and decent work for all; (6) Goal 15: protect, restore and promote sustainable use of terrestrial ecosystems, sustainably manage forests, combat desertification and halt and reverse land degradation and halt biodiversity loss.

We recently designed an innovative, multi-actor approach based on the integration of a variety of data, skills, expertise and competences to create a knowledgeable platform and a strategic plan for the sustainable consumption and production of healthy, identifiable, safe and traditional food products. This project, with the SHAPE acronym standing for Sustainable, Healthy and identifiable Agri-food Production Enterprise, represents the natural evolution of the PASSI project, as it is molded upon a consumer-centered food system model that comprises individual taste and preferences; equitable access and affordability; socio-economic, psychosocial and behavioral determinants of food behaviors; cultural identity; innovative marketing strategies; and comprehensive nutrition education all with the goal of strengthening the valorization of the health benefits of the Mediterranean diet in different populations groups. This innovative model could be developed and implemented in a series of "Mediterranean" countries, including Italy, Greece, Malta, Lebanon, Jordan (partners of the project), Portugal and Spain, where the dramatic changes that have occurred in both food systems and consumer eating habits and behavior in recent decades have led to a significant increase in the incidence rates of NCDs, along with high percentages of overweight and/or obese children. In this respect, the innovative SHAPE approach could also be used, in the short and medium term, as a primary prevention strategy to tackle the increasingly large prevalence of noncommunicable diseases in Mediterranean regions.

Importantly, food environments should be transformed, before anything, culturally, by defining and implementing measures of nutritional education, from primary schools to adulthood, to seminally diffuse healthy behaviors not only in term of eating patterns but also regarding physical activity.

In this respect, the Department of Health (DASOE) of the Sicilian Region initiated the Program named "Formazione Educazione Dieta" (FED), aimed at the promotion and diffusion of healthy lifestyles, according to the Mediterranean Diet (Requirez et al. 2016). The major methodological strength of the FED program consists of an intervention of health prevention and promotion based on two strategic elements: a centralized planning, adjustable according to outcome indicators, and a multiyear action plan, with cyclic activities throughout three regional networks (health, education and agribusiness). The program is synergistically focused on: (a) a cascade training program, aimed at qualifying people to become expert trainers/educators able to influence the cultural changes, behaviors and lifestyles of the population, by acting specifically on the various recipients, in accordance with appropriate methodologies and evidence-based content; (b) the activation of local networks built to promote capillary activities in the regional territory (see Figure 8).

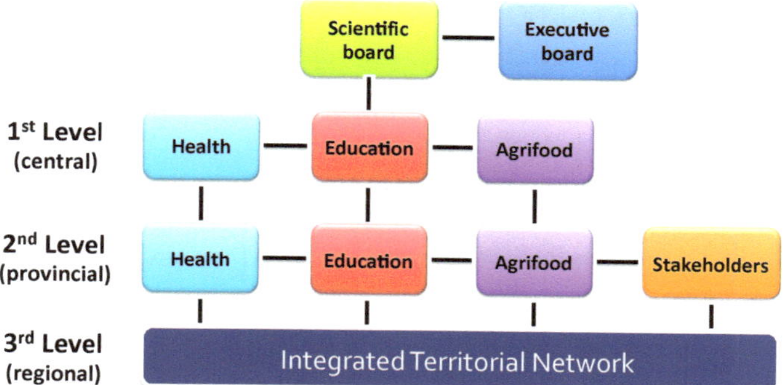

Figure 8. The FED operational model. The program is based on a cascade training system, governed by the Scientific Board and divided into 3 levels: (a) central, addressed to beneficiaries from Health, Education and Agrifood macroareas (1st level trainers); (b) provincial, addressed to beneficiaries from selected macroareas and stakeholders (2nd level trainers); (c) regional, with the creation of an Integrated Territorial Network consisting of representatives from the Scientific Board, 1st and 2nd level trainers for seminal diffusion of education activities throughout the region. Source: Figure by the author.

Using an integrated multiprofessional approach, the experts develop an operational and training program with centrality and uniqueness, making it uniformly applicable to the distinct organizational realities of the Sicilian territory.

Today, just ten multinational companies control over 70% of the food supply on the planet, owning nearly 500 different brands and providing the largest market ever of unhealthy, high-calorie food. Against their economic and organizational supremacy, governments should put in place strict policies to protect the production and marketing of traditional food through investments in short-chained, territorial healthy products.

In this framework, outstanding importance is also assigned to the transition from a linear economy to a circular economy in the agrifood industry, providing an array of opportunities at all stages, from primary production using precision agriculture techniques, to the recycling and use of agricultural byproducts or waste to produce functionalized, healthy food.

In a holistic view, global efforts are being made through several international initiatives, notably the 2030 Agenda for Sustainable Development including its 17 Sustainable Development Goals (SDGs), to radically modify food systems and environments with the ambitious objective of preventing NCDs while protecting the human right of equitable, accessible and affordable nutritious food.

Acknowledgments: The DiMeSa project was financed by a grant (PON02 00667-PON02 00451 3361785) from the Italian Ministry of University and Research (MIUR) to the The AgroBioPesca Technology Cluster. The FED Regional Programme was officially adopted by the Sicilian Region with the 1778/13 decree of the Regional Ministry of Health on 30 December 2013. The author is grateful to both Mario Enea (University representative) and Ing. Antonio Giallanza (Project Manager), from the Department of Engineering of Palermo University, for their invaluable contribution to the DiMeSa and 123 PASSI projects. I also wish to thank Tiziano Caruso and Giuseppe Di Miceli, from the Department of Agriculture and Forestry Sciences of Palermo University, for their precious collaborative efforts, respectively, in the production of monocultivar EVOs and the OFI-supplemented pasta in the DiMeSa study, and last but not least, Nicola Locorotondo for his continuous enthusiastic support. The author wishes to acknowledge the invaluable work conducted by friends and colleagues of the FED Scientific Board, including Elena Alonzo, Pietro Di Fiore, Daniela Falconeri, Francesco Leonardi and Salvatore Requirez, for the implementation of the FED Regional Programme.

Conflicts of Interest: The author declares no conflict of interest.

References

Branca, Francesco, Anna Lartey, Stineke Oenema, Victor Aguayo, Gunhild A. Stordalen, Ruth Richardson, Mario Arvelo, and Ashkan Afshin. 2019. Transforming the food system to fight non-communicable diseases. *BMJ* 364: 1296. [CrossRef] [PubMed]

Carruba, Giuseppe, Letizia Cocciadiferro, Antonietta Di Cristina, Orazia M. Granata, Cecilia Dolcemascolo, Ildegarda Campisi, Maurizio Zarcone, Maria Cinquegrani, and Adele Traina. 2016. Nutrition, aging and cancer: Lessons from dietary intervention studies. *Immun Ageing* 13. [CrossRef]

Dahlgren Göran, and Margaret Whitehead. 1991. *Policies and Strategies to Promote Social Equity in Health*. Stockholm, Sweden: Institute for Futures Studies.

Gale, Edwin A. M. 2002. The rise of childhood type 1 diabetes in the 20th century. *Diabetes* 51: 3353–61. [CrossRef] [PubMed]

Giampaoli, Simona, Vittorio Krogh, Sara Grioni, Luigi Palmieri, Massimo M. Gulizia, Jeremiah Stamler, and Diego Vanuzzo. 2015. Eating behaviours of Italian adults: Results of the osservatorio epidemiologico Cardiovascolare/Health Examination Survey. *Epidemiologia e Prevenzione* 39: 373–79. [PubMed]

Greendex. 2014. *Consumer Choice and the Environment—A Worldwide Tracking Survey—Full Report*. National Geographic/GlobeScan. Available online: https://globescan.com/wp-content/uploads/2017/07/Greendex_2014_Full_Report_NationalGeographic_GlobeScan.pdf (accessed on 24 April 2021).

Grunert, Klaus G. 2005. Food quality and safety: Consumer perception and demand. *European Review of Agricultural Economics* 32: 369–91. [CrossRef]

High Level Panel of Experts on Food Security and Nutrition. 2017. *Nutrition and Food Systems. A Report by the High Level Panel of Experts on Food Security and Nutrition of the Committee on world Food Security, Rome*. Available online: http://www.fao.org/3/i7846e/i7846e.pdf (accessed on 24 April 2021).

Hoefner, Daniel S. D. Shannon Hodel, John F. O'Brien, Earl L. Branum, Deborah Sun, Irene Meissner, and Joseph P. McConnell. 2001. Development of a rapid, quantitative method for LDL subfractionation with use of the quantimetrix lipoprint system. *Clinical Chemistry* 47: 266–74. [CrossRef] [PubMed]

Hunter, David J., and K. Srinath Reddy. 2013. Noncommunicable diseases. *The New England Journal of Medicine* 369: 1336–43. [CrossRef] [PubMed]

INEA. 2013. *L'agricoltura nella Sicilia in Cifre 2012*. Poznan: INEA.

ISTAT. 2010. *6° Censimento generale dell'agricoltura*. Chicago: ISTAT.

Kacen, Jacqueline J., and Julie A. Lee. 2002. The influence of culture on consumer impulsive buying behavior. *The Journal of Consumer Psychology* 12: 163–76. [CrossRef]

Pisani, Paola, Fabrizio Faggiano, Vittorio Krogh, Domenico Palli, Paolo Vineis, and Franco Berrino. 1997. Relative validity and reproducibility of a food frequency dietary questionnaire for use in the Italian EPIC centres. *International Journal of Epidemiology* 26 S1: S152–S160. [CrossRef] [PubMed]

ReNCaM. 2014. *Registro Nominativo delle Cause di Morte in Sicilia*. Palermo: Assessorato Regionale Salute.

Requirez, Salvatore, Francesco Leonardi, Giuseppe Carruba, Pietro Di Fiore, Daniela Falconeri, and Elena Alonzo. 2016. Program "Formazione Educazione Dieta" (FED) aimed at the promotion of Mediterranean Diet. *Sistema Salute* 60: 51–67.

Rizzo, Manfredi, and Kaspar Berneis. 2006. Should we measure routinely the LDL peak particle size? *International Journal of Cardiology* 107: 166–70. [CrossRef] [PubMed]

Turner, Christopher, Anju Aggarwalb, Helen Wallsc, Anna Herforthd, Adam Drewnowskie, Jennifer Coatesf, Sofia Kalamatianoua, and Suneetha Kadiyalaa. 2018. Concepts and critical perspectives for food environment research: A global framework with implications for action in low- and middle-income countries. *Global Food Security* 18: 93–101. [CrossRef]

World Health Organization. 2013. *Global Action Plan for the Prevention and Control of Noncommunicable Diseases 2013–2020*. Geneva: World Health Organization.

World Health Organization. 2018a. *World Health Statistics 2018—Monitoring Health for the SDGs, Sustainable Development Goals*. Geneva: World Health Organization.

World Health Organization. 2018b. *Global Health Estimates 2016: Deaths by Cause, Age, Sex, by Country and by Region, 2000–2016*. Geneva: World Health Organization, Available online: http://www.who.int/healthinfo/global_burden_disease/estimates/en/index1.html (accessed on 24 April 2021).

Wijnhoven, Trudy M. A., Joop M. A. van Raaij, Angela Spinelli, Gregor Starc, Maria Hassapidou, Igor Spiroski, Harry Rutter, Éva Martos, Ana I Rito, Ragnhild Hovengen, and et al. 2014. WHO European Childhood Obesity Surveillance Initiative: Body mass index and level of overweight among 6–9-year-old children from school year 2007/2008 to school year 2009/2010. *BMC Public Health* 14: 806. [CrossRef] [PubMed]

Impact of Public Health and Sustainability of Global Health Action for Achieving SDG 3

Florian Fischer and Franziska Carow

1. Introduction

In current debates—be it on political, societal or economic levels—the term "globalisation" is used quite frequently. However, it tends to be misused and overused, because there is missing clarity about its meaning and its impact on individuals and society. Despite the complexity behind it, globalisation can be summarised as a social transformation, manifesting in spatial (e.g., migration, as well as exchange of information, capital, goods and serves across national borders), temporal (e.g., caused by speeded up interaction through modern communication and transportation technologies) (Oke 2009; Lee 2004), and cognitive changes—meaning the way how we see ourselves and the world surrounding us. This leads to changes in creating, exchanging and applying knowledge, ideas, norms, beliefs, values, cultural identities, and other thought processes (Lee 2003, 2004; Bettcher and Lee 2002). Currently, humanity is facing a rapid global change, which differs substantially in its extent from previous episodes of global change (Sachs et al. 2019; National Academies of Sciences, Engineering and Medicine 1992). Due to demographic, epidemiological, environmental, social, political, economic, technological, and cultural driving forces, we have encountered a "transformative change" which is gaining momentum (Sachs et al. 2019). All of these forces are interconnected, leading to the overall complexity associated with changes in the health status of individuals and (sub-)populations. Health impacts of globalisation are positive and negative at the same time. The impact may vary according to geographical location, sex/gender, age, ethnic origin, education level, and socioeconomic status (Lee 2004; Huynen et al. 2005).

For that reason, globalisation causes divergence as it can be seen in several examples such as the global reorganisation of production and the emergence of a global labour market, the incremental mobility of financial capital, the growing importance of binding trade agreements and processes as well as global political agendas focusing on sustainability, and the persistence of debt crises particularly in low- and middle-income countries (Schrecker et al. 2008). All of these examples are not directly from the health(care) sector, but all of them impact on health. Therefore, reducing global burden of disease and improving health is an effort, which goes much beyond healthcare system and health policies. It needs to include policymakers and stakeholders at community, regional, country, and global level, originating from all disciplines and policy fields. This relates to the "Health in all policies" approach, which determines the impacts of policymaking of all kinds and of all levels on health and the health system. This approach supports policymakers to include health in their decisions at the regional and national levels, but can also be adapted for supranational level decision-making (WHO 2014).

The United Nations declare health and well-being as a specific goal of their overall 17 worldwide Sustainable Development Goals (SDGs) (UN 2015b). In terms of ensuring healthy lives and promoting well-being for all at all ages—as claimed in SDG 3—this approach should be transferred to the global level. The targets and indicators for SDG 3 can be summarised as a call for universal health coverage, including quality essential access to health services, to sexual and reproductive healthcare services, to safe and affordable medicine and vaccines for everyone, the reduction of the global neonatal and maternal mortality, as well as reduction of suffering and deaths caused by communicable and non-communicable diseases, pollution and accidents (UN 2016). For reaching these aims and allowing for a sustainable development, an acknowledgement of sectors others than public health, e.g., food and agriculture industry and education, is crucial (Laaser and Epstein 2010). Furthermore, this emphasises the need to consider the aims of all other SDGs also in the context of their relationship to SDG 3, in terms of direct or indirect effects on health.

Impacting on the health status of the world population as well as allowing sustainability in the context of public health and global health are challenging tasks. Within this contribution, we describe the challenges for public health on the global scale. This is illustrated by four case studies, serving as examples for the relevance of sustainability-focussed global health action for achieving SDG 3. This contribution provides overarching suggestions and policy advice for further improvements to strengthen a global response on today's challenges and to gain the targets set within the SDGs.

2. Global Health

Global health is defined as "an area for study, research, and practice that places a priority on improving health and achieving equity in health for all people worldwide" (Koplan et al. 2009). Beaglehole and Bonita (2010) emphasise the aspect of transnationality of research and action. The aim of global health is to overcome the national–international divide and the divide between countries of different income levels through mutual exchange and partnership. This includes joint responsibility, funding, agenda setting, planning, and implementation (Berner-Rodoreda et al. 2019; Gautier et al. 2018; Koplan et al. 2009). This characterisation is at least partly in line with a dialectic approach for defining global health. According to this, "global" can be understood as "worldwide", as "supraterritorial", as "transcending national boundaries", and as "holistic" (Bozorgmehr 2010). Therefore, global health has a definitive political dimension. It corresponds with the aims of the SDGs for providing universal health coverage and leaving no one behind (UN 2015b) and, therefore, aims to reduce health inequities (in terms of unfair, avoidable differences arising from poor governance, corruption or social exclusion) and health inequality (such as the uneven distribution of health or health resources in or between populations) (Reidpath and Allotey 2007).

2.1. Evolution of Global Health

Global health has experienced a major and rapid development in the past years (Martin et al. 2014). In general, global health is based on concepts, aims, and methods of public health. It followed similar foci during its evolution, and has, therefore, been focussing mainly on disease prevention (particularly infectious diseases)—such as "old" public health"—by providing access to clean water and more nutritious food, by promoting hygiene (e.g., due to body and hand washing, disinfection, town sewerage systems, quarantine laws), as well as by implementing population-based surveillance and screening mechanisms (Kellehear 2017). The evolution of global health went on to follow the ideas of "new" public health, which takes a holistic and interdisciplinary perspective on aspects related to health and well-being. Modern public health is understood as "the science and art of preventing disease, prolonging life and promoting health through the organised efforts and informed choices of society, organisations, public and private, communities and individuals" (Wanless 2004, p. 3). This definition includes health promotion, prevention, treatment and rehabilitation of disease, and the efforts of all stakeholders. Inter alia, public health genuinely deals with the reduction of health inequalities and includes social determinants of health—not only by prevention on a behavioural, but also on a structural level. Furthermore, public and global health have a focus on health policy and governance (Carlson et al. 2015; Barbazza and Tello 2014). The scope has widened because damage to the environment can also impact on health, as seen in the effects of climate change, environmental degradation, and biodiversity loss (Roe 2019). For that reason, one can understand health as part of each of the 17 SDGs.

2.2. Health as a Human Right and Global Imperative

In the preamble of the World Health Organization's (WHO) constitution, health was declared as "a state of complete physical, mental and social well-being and not merely the absence of disease or infirmity" (WHO 1946). In 1948, the Universal Declaration of Human Rights pointed out that health is part of the right to an adequate standard of living (art. 25) and was reaffirmed a human right in the Declaration of Alma-Ata in 1978 ("Health for All"), which emerged as a central milestone in the field of public health, is it expresses "the need for urgent action by all governments, all health and development workers, and the world community to protect and promote the health of all the people of the world" (WHO 1978). The 40th anniversary of this declaration was used to renew its focus on the role of primary healthcare under the lens of universal health coverage and the SDGs in the Declaration of Astana (Walraven 2019; Tilford 2018; WHO 2018a).

The fact that there has been a legal basis for the right to health for over seventy years, and that the goal of universal health coverage and equality have thus been political issues for such a long period of time, is accompanied by ethical considerations. This was particularly evident during the HIV/AIDS crisis starting in the 1980s, with the following movement against the patent system for HIV/AIDS medication showing how strongly human rights, ethics and health are related. The disparate burden of HIV in low-income countries in conjunction with the AIDS activist movement have led to a consideration of health as a global imperative over the past three decades, and revived a human rights-based approach to healthcare (Keusch et al. 2010). The activities of societies and stakeholders worldwide put enough pressure on the system to decrease costs for HIV/AIDS treatment by up to 99% (Meier et al. 2018). Since this period, an increasing number of countries have started to implement operating systems and policies that centre on health as a human right (Gostin et al. 2018).

3. Challenges for Global Health Action

Despite the progress that has been made in global health in the past decades, several manifold and complex challenges are hindering the success of developing, implementing and evaluating effective and sustainable action plans to improve the world population's health. For example, Ridde (2016) emphasised the need for more and better implementation science in global health, for bridging the "know–do gap" (Means et al. 2016). A common mantra in implementation science in public health and global health is that we know what to do, but not how to do it: "Therefore, it is not enough to know if a health intervention is effective; it is also necessary to understand why the intervention works, how, for whom and in which contexts" (Ridde 2016, p. 1). Impact evaluations need to go hand in hand with the evaluation of implementation. We need further evidence to understand implementation processes, causal mechanisms, and contextual factors affecting the outcomes of complex interventions in global health (Ridde et al. 2020). This has to go along with a strong theoretical foundation (Ridde et al. 2020; Van Belle et al. 2017). Until now, implementation research in low- and middle-income countries has mainly focused on evaluating the effects of implementation strategies. Problems of scale-up and sustainability, which are key issues for global health interventions, have not adequately been addressed (Alonge et al. 2019).

In addition to the lack of implementation research, another major challenge is to consider the diversity and heterogeneity in the needs of the many economically, socially, culturally, climatically and demographically different regions and subpopulations (Mason et al. 2017). Furthermore, there is not enough or only fragmentary research and limited data availability and/or quality in many countries, which additionally compounds estimating problems and needs on various levels (Asma et al. 2019). However, several attempts try to provide an overview in terms of problems and successes of global health as well as outline the progress in reaching SDG 3. Reporting on the global burden of disease and its development over time and space and in relation to the SDGs is crucial, as performed by the Global Burden of Disease study (Lozano et al. 2018), the United Nations (UN 2019b) and WHO (WHO 2019b).

3.1. Financing

In the past, global health, especially in industrialised countries, was seen from a perspective of providing development aid. Development projects have frequently been implemented with a vertical structure. However, these top-down approaches used by several funders often had no (long-term) effects. Due to a lack of coordination, inefficient use of funding, and a strong focus on improving individual health within these approaches and programs directed at specific diseases—rather than public health or community approaches—many projects had no sustainable denouement and sometimes even led to adverse effects on the health status of the addressed country or region (Laaser and Epstein 2010; McCoy et al. 2009; Chang et al. 2019).

Despite the overall steadily increasing amounts of money used for global health financing, one major reason for the only modest achievements since the Declaration of

Alma-Ata is, therefore, its ineffective use (Laaser and Epstein 2010). The WHO points out that the global needs are clearly not being met by current funding. Although external funding (aid) represents less than 1% of global health expenditure (WHO 2019b), only 0.3% of all direct grants were received by low-income countries in 2016 (WHO 2018b), with an increasing trend for low-income countries and a declining proportion of health spending in middle-income countries (WHO 2019b). At the same time, the funding situation of the official development assistance (ODA), focussing on the poorest countries worldwide, has increased by 61% in real terms from 2010 to 2019 (UN 2019a). Nevertheless, the rural poor of large middle-income countries are neglected, although they are important target groups. It is important to note that out-of-pocket spending especially affects the poor. In 2010, 11.7% of the global population spent at least 10% of their household budget on healthcare, an estimated 1.4% worldwide were impoverished by financing healthcare the same year (WHO 2018b). This inequality strongly addresses SDG 3, but also SDG 1 ("No poverty"), SDG 2 ("Zero hunger"), and SDG 10 ("Reduced inequalities"). For achieving the goal of universal health coverage and better health and well-being, an efficient and effective financing is fundamental.

Furthermore, the United Nations and WHO state that workforce to provide health services in many countries is not strong enough (WHO 2019b; UN 2019a). Within the last five years, 40% of all countries had fewer than ten medical doctors per 10,000 people (UN 2019a). The education of medical staff is crucial to implementing essential healthcare and to covering the needs of low- and middle-income countries. This problem not only addresses SDG 3, but also SDG 4 ("Adequate education"), SDG 10 ("Reduced inequalities"), SDG 11 ("Sustainable cities and communities") and SDG 16 ("Peace, justice and strong institutions").

All these aspects highlight the necessity for re-organising the ODA (Chan 2013), as "at least half of the world's population does not have full coverage for essential health services" (WHO 2018b). Additionally, a review and modelling of the past, present and future financing of global health, for the years 1995 to 2050, projects a persistent and growing gap in per capita health spending, leading to growing disparities between countries worldwide (Chang et al. 2019).

3.2. Responsibility

In 2019, the WHO designated leadership and engaging partnership on health-matters, the setting of standards and monitoring their implementation, shaping the research agenda, assessing health trends and articulating policy options as its core-functions (WHO 2019a). Although many countries support the WHO and its goals, the financing of the WHO is a problem that needs to be addressed. WHO continues to experience immense financial stress, as it is currently visible in public debates after the United States of America called to suspend their funds (Nature 2020). This leads to questions about the future of WHO in global health governance (Reddy et al. 2018). The WHO is funded by mandatory and voluntary member state contributions and other non-state organisations (Clift and Røttingen 2018). Clift and Røttingen (2018) have stated that a shift to tied voluntary donations can be observed within the last few years. This means that the WHO is not in power of controlling how 80% of its budget is spent. Bennett et al. (2018) also claim that 80% of the funding of the WHO is earmarked for specific purposes. They emphasise that many countries resist non-earmarked funding, because funders want to know where the money goes and need to justify their donations. However, a focus on global priorities (i.e., non-communicable diseases and universal health coverage), which currently cannot be given enough attention due to funding, requires a more transparent, flexible, and predictable fund allocation for the WHO (Bennett et al. 2018; Reddy et al. 2018).

Progress has been achieved in terms of responsibility, when the Millennium Development Goals (UN 2015a) were expanded by the amendment of sustainability in 2015. To accelerate this progress, 193 countries adopted the SDG Agenda 2030 at the Sustainable Development Summit in New York, in September 2019. Additionally, countries and stakeholders can voluntarily undertake acceleration actions that "contribute to a speeded up implementation of the 2030 Agenda" (UN 2019b). The rationale for international cooperation is most compelling in the field of global health, since global interdependence is most acute. Coherence across all areas of public policy is needed to realise health equity and well-being for all. Governance mechanisms and intersectoral initiatives are needed and should be based on the SDGs, which provide a framework for strengthening policy coherence for achieving health equity (WHO Europe 2019).

During the coronavirus pandemic the crucial need for scientific collaborations in both the public and private sectors at the global scale became visible to develop diagnostics, vaccines and treatments in order to tackle health emergencies. However, this was also the time when we have seen various national responses, with a lack of globally coordinated approaches and missing responsibility for global action. Furthermore, the collateral effects of the COVID-19 pandemic caused by global economic downturn, social isolation and movement restrictions, are unequally distributed and mainly affecting those in the lowest power strata of societies (Shadmi et al. 2020).

3.3. Changing Patterns of Morbidity, Mortality, and Population Dynamics

Global health actions need to encounter the demographic (Kirk 1996) and epidemiological transition (Omran 1971). The demographic transition describes a population's shift from high mortality and high fertility to low mortality and low fertility. However, this pattern in population dynamics does not happen uniformly across countries and regions worldwide. Therefore, some countries experience demographic ageing and population decline (particularly industrialised countries), whereas others are characterised by a robust population growth (Blue and Espenshade 2011). Recent conceptual adjustments incorporated the nonlinear changes towards very low fertility and a diversity of union and family types (Zaidi and Morgan 2017). Despite these considerations, one needs to recognise that the demographic patterns affect population's health. The epidemiological transition describes the shift from infectious diseases to non-communicable disease entities. However, this shift takes place at different pace and from different starting points (Jamison et al. 2013).

For example, neonatal, infant, and child mortality is still an important (global) public health challenge in low- and in some middle-income countries (Burstein et al. 2019). The same applies to infectious disease. Infectious diseases are a major risk in low- and middle-income countries due to poverty, insufficient healthcare, unawareness or unavailability of preventive measures (Dye 2014), and the impact of climate change additionally facilitate outbreaks (Liang and Gong 2017). Vaccination is known as one of the most cost-effective and successful public health interventions with enormous contributions to global health (Greenwood 2014). A significant proportion of childhood mortality in low-income countries has been reduced due to vaccination programs which promote herd immunity, eradicate diseases in the long-term and by doing so ensure health and well-being (Glatman-Freedman and Nichols 2012). For example, vaccination coverage for the prevention of diphtheria, tetanus and pertussis increased from 72% in 2000 to 85% in 2015 (UN 2019a) as a result of the Global Vaccine Action Plan (WHO 2013) and the Global Alliance for Vaccines and Immunisation (GAVI), a public–private partnership committed to saving children's lives and protecting people's health by increasing access to vaccinations (GAVI 2019).

In terms of global health—oriented toward the concepts of "old" public health—the Ebola (Honigsbaum 2017) and Zika (McNeil and Shetty 2017) outbreaks were obvious signals highlighting the need for pandemic preparedness for avoiding public health emergencies of international concern. Experiences gained during the global health crisis triggered by Ebola led to improvements in early response systems, in the development of interdisciplinary and intersectoral responses, and, most importantly, in close cooperation between civil society and communities (Raguin and Girard 2018). The perspective of "new" public health also takes social determinants into account: for example, cholera is a strong indicator of inequality and lack of social and economic development, because it disproportionately affects the world's poorest and most vulnerable populations. Most of the time, cholera outbreaks occur as a result of conflicts, natural disasters, famines, unsafe drinking water and deficient sanitation facilities (WHO 2018b). Cholera treatment requires early detection and immediate healthcare and medication. For cholera outbreaks, the WHO recommends the establishment of surveillance systems and rapid response teams and calls for supply readiness and higher laboratory capacities (WHO 2018b). SDG 3 is not the only goal pertinent to the issue of preventing or eliminating infectious diseases: reducing inequalities (SDG 10); ensuring the availability and sustainable management of water and sanitation (SDG 6); protecting, restoring and promoting sustainable use of ecosystems, and halting biodiversity loss (SDG 15); taking action to combat climate change and its impact (SDG 13); promoting peaceful and inclusive societies for sustainable development; providing access to justice for all and building effective, accountable and inclusive institutions at all levels (SDG 16) also relate to this issue.

Furthermore, the global burden of non-communicable diseases is no longer just a phenomenon of high-income countries. Several non-communicable diseases are socially patterned and related to behavioural risk factors such as tobacco smoking, alcohol consumption, low levels of physical activity, and unhealthy eating habits, which are getting much more common also in low- and middle-income countries (Marmot and Bell 2019; Stringhini and Bovet 2017). These countries are prone to suffer from a double burden of disease due to a high proportion of infectious diseases and non-communicable diseases (Boutayeb 2006). For example, in 2016 almost 340 million (18.4%) children and adolescents (5–19 years) were overweight or obese globally and, thus, at a high risk for suffering from non-communicable diseases later in life. The rate of obese children is higher in high-income countries, but the number of obese children and adolescents in low- and middle-income countries is increasing faster (WHO 2018b).

4. Case Studies: Global Health and Sustainability

Sustainability has become a central criterion in evaluating public health programmes. In global health, a major shift took place from development to sustainability as well (The Lancet 2012). However, there are discussions about the conceptualisation and measurement of sustainability criteria. Overall, health is an investment that is itself sustaining and sustainable (Yang et al. 2010).

Global health action requires a mind shift towards a new political and social movement for health (Kickbusch 2014) which addresses the social, cultural, physical, environmental, commercial and political determinants of health. These approaches need to ensure a balance between domestic and global action while recognising the commitment to common goals such as the SDGs (Kickbusch 2016). Politicians and scientists representing a wide range of disciplines, businesses, civil society, and local communities, need to be new agents of change (Moallemi et al. 2019). A global and holistic perspective is needed for understanding (transnational) health issues and its determinants. This perspective needs to be applied at the local level to improve health and to gain sustainability (Rowthorn 2015). For highlighting the complexity of gaining impact and sustainability in global health and for showing the need of participatory, inter- and transdisciplinary approaches, some examples in form of case studies are delineated.

4.1. Antimicrobial Resistance

The rapid and ongoing spread of antimicrobial resistance poses a serious threat to global health (Watkins and Bonomo 2016; Hay et al. 2018). The indiscriminate use of antibiotics in human medicine and agriculture are a main driver contributing to resistance (Laxminarayan et al. 2013), leading to health crises arising from infections that were once easy to treat (Hay et al. 2018). In high-income countries, patients with resistant infections frequently have the opportunity to turn to newer-generation antibiotics, which are more expensive. In low-income countries, where infectious diseases are leading to a high disease burden, patients might be unable to obtain or to afford second-line treatments (Laxminarayan et al. 2013). Infections resistant to antimicrobial treatment frequently result in longer hospital stays, higher healthcare expenses, and increased mortality (Hay et al. 2018).

To combat antimicrobial resistance, comprehensive national and international plans, like the Antibiotic Stewardship programme in the European Union (Allerberger et al. 2009), are needed to allow for rationale antibiotic use in hospitals (Laxminarayan et al. 2013). Furthermore, due to the connection between antimicrobial resistances and the agricultural sector, the One Health approach comes into play. The One Health approach supports global health security by improving coordination, collaboration and communication at the human-animal-environment interface. For doing so, multiple disciplines working locally, nationally, and globally are included, to address shared health threats such as zoonotic diseases, antimicrobial resistance, food safety and others (American Veterinary Medical Association 2008; Sinclair 2019).

4.2. Climate Change

The 2019 report of The Lancet Countdown on health and climate change emphasised that the health of a child born today will be affected by climate change over its whole lifespan (Watts et al. 2019). Therefore, climate change is an emerging threat to global health, as it has myriad implications for the health of humans and the ecosystems (Machalaba et al. 2015; Patz et al. 2014). The effects of climate change are closely linked to social and ecological determinants of disease mitigating or exacerbating forecasted adverse health outcomes (Machalaba et al. 2015). Climate change is also highly inequitable, as the greatest risks are to the poorest populations, who have contributed least to greenhouse gas emissions (Campbell-Lendrum and Corvalán 2007). There is a substantial overlap between underlying determinants of health inequity and environmental change. Friel et al. (2008) claimed that they "are signs of an economic system predicated on asymmetric growth and competition, shaped by market forces that mostly disregard health and environmental consequences rather than by values of fairness and support". For that reason, multidisciplinary collaboration and a shift in priorities in economic development towards healthy forms of urbanisation and infrastructure, more efficient and renewable energy sources, and a sustainable and fairer food system is needed (Friel et al. 2008). Without immediate actions, climate change will impact on the health of current and future generations, will pose challenges to already overwhelmed health systems, and will undermine the progress towards achieving the SDGs and universal health coverage (Watts et al. 2019).

4.3. Migration

Globalisation is characterised by increases in population movement. International migration is a complex phenomenon affecting a multiplicity of economic, social and security aspects. However, with about 250 million international migrants in 2015, and a projection of more than 400 million in 2050 (IOM 2018), and significantly more people moving within their country of birth, there is an urgent need to engage with the topic of migration in global health (Wickramage et al. 2018). Therefore, the UCL-Lancet Commission on Migration and Health called on "nation states, multilateral agencies, non-governmental organisations, and civil society to positively and effectively address the health of migrants by improving leadership and accountability" (Abubakar et al. 2018). Despite the human right to health, national sovereignty concerns frequently overshadow legal norms, as attention to migration focuses largely on security concerns (Abubakar et al. 2018). This has been highly visible at the so-called "refugee crisis" impacting Europe in 2015/2016, which was much more a "crisis of solidarity" (Bozorgmehr and Wahedi 2017). In this regard, fundamental human rights have been restricted to asylum seekers in many recipient countries, because access to adequate healthcare was denied (Bozorgmehr and Razum 2016).

In the context of global health, it is mandatory to consider the "Health in all policies" approach, because the steady increase in international migration has led to hostile migration policies worldwide, such as expanded border controls in the European Union and attempts to rescind legal protection granted to undocumented migrants in the United States. All these policies have not been designed to negatively affect migrant's health, but their role as social or political determinants of health is undisputable (Juárez et al. 2019).

4.4. Digitalisation

Digitalisation, one of the megatrends of our time, is of particular relevance for global health. First of all, it affects our lives like nearly no other societal, technological or economic development. Second, digitalisation is global itself. In developed and developing countries alike, modern communications technology is no longer a convenience—it is a necessity. Due to its importance, the increase in access to information and communication technologies is addressed in SDG target 9c. Digital strategies have been recognised as a critical strategy for health systems strengthening to help meet the SDGs and universal health coverage targets (Labrique et al. 2018). Digitalisation is crucial for precision global health, which is describes as an approach similar to precision medicine. Through innovation and technology, it facilitates better targeting—or even tailoring—of public health interventions on a global scale (Flahault et al. 2017). For that reason, digitalisation brings new potential for global health, but further reflections on its ethical implications and social impacts are needed (Dockweiler and Fischer 2019).

5. Conclusions

Never before has global change happened so quickly. The societal, epidemiological and demographic changes we face require new and particularly global strategies to gain the common goal of universal health coverage. Global health is closely linked to sustainable actions. Indeed, global health actions need to be sustainable to improve the world population's health in the long term and to reduce health inequities.

All global health actions should be based on the best available scientific evidence. Therefore, Rudan and Sridhar (2016) summarised the "basic needs" of the global health research system that emerged from the past:

- coordination of funding;
- prioritisation of the plentiful research ideas;
- recognition of results of successful research;
- broad and rapid dissemination of results and their accessibility;
- evaluation of return on investments.

Best practice examples could serve as prototypes for a better coordination and organisation for the common goal of a higher standard in public and global health without causing substantial or additional costs for research, development and design (Rudan and Sridhar 2016).

Until the recent past, the European Union for example defined global health mainly in terms of strengthening "global and third countries' national capacities of early prediction, detection and response to global health threats" (European Commission 2010) rather than focusing on cross-border health threats. Germany used its presidencies of the G7 and G20 summits in 2015 and 2017 to give more prominence to global health in supranational political discussion (Berner-Rodoreda et al. 2019). However, the focus has been relatively narrow. Therefore, McBride et al. (2019) recommended expanding the focus to neglected SDG3 health targets to place greater emphasis on upstream determinants of health, provide stronger commitment to equity and leaving no-one behind, adopt explicit commitment to rights-based approaches, and make commitments that are of higher quality and which include time-bound quantitative targets and clear accountability mechanisms.

Reaching SDG 3 targets is undeniably an enormous challenge that comes along with structural and financial demands, which are compounded by additional hazards, such as political conflicts, natural disasters and famines, but also new global problems, for example antibiotic resistances and the adverse health effects causes by climate change. Therefore, sustainable development needs to be at the core of the global agenda (Kickbusch 2014), allowing for a long-term implementation of public health infrastructure, is needed to accomplish SDG 3. Action on the social determinants of health—based on a "Health in all policies" approach—is required to reduce inequities in health (Donkin et al. 2018). Accountability, vested interests, ethics and democratic legitimacy are conditional for future sustainability of population health (Byskov et al. 2019).

The overall changes in the past years, in terms of the increasingly globalised nature of economy, society and culture, combined with, e.g., the effects of climate change and environmental degradation as well as the evolution of antibiotic resistance, have led to shift the boundaries. These factors expose both new and forgotten similarities between populations. Furthermore, they highlight the need for global cooperative responses to health threats. Therefore, the grand challenges can serve as "a catalyst for global solidarity, which justifies, and provides motivation for, the establishment of solidaristic, cooperative global health infrastructures" (West-Oram and Buyx 2017, p. 212).

Reaching the goal of a better health and universal health coverage also implies changes in other areas than health and, thus, addresses more SDGs than just SDG 3. Therefore, integration of global health concerns into the law and governance of other, related disciplines should be given high priority. This emphasises the need for developing and implementing a "global health" policy, and not only a global "health policy".

Author Contributions: Both authors jointly conceptualised and wrote this contribution.

Funding: This research received no external funding.

Conflicts of Interest: The authors declare no conflict of interest.

References

Abubakar, Ibrahim, Robert W. Aldridge, Delan Devakumar, Miriam Orcutt, Rachel Burns, Mauricio L. Barreto, Poonam Dhavan, Fouad M. Fouad, Nora Groce, Yan Guo, and et al. 2018. The UCL–Lancet Commission on Migration and Health: The Health of a World on the Move. *The Lancet* 392: 2606–54. [CrossRef]

Allerberger, Franz, Roland Gareis, Vlastimil Jindrák, and Marc J. Struelens. 2009. Antibiotic Stewardship Implementation in the EU: The Way Forward. *Expert Review of Anti-Infective Therapy* 7: 1175–83. [CrossRef]

Alonge, Olakunle, Daniela Christina Rodriguez, Neal Brandes, Elvin Geng, Ludovic Reveiz, and David H. Peters. 2019. How is Implementation Research Applied to Advance Health in Low-Income and Middle-Income Countries? *BMJ Global Health* 4: 001257. [CrossRef]

American Veterinary Medical Association. 2008. *One Health: A New Professional Imperative.* Schaumburg: American Veterinary Medical Association.

Asma, Samira, Rafael Lozano, Somnath Chatterji, Soumya Swaminathan, Maria de Fátima Marinho, Naoko Yamamoto, Elena Varavikova, Awoke Misganaw, Michael Ryan, Lalit Dandona, and et al. 2019. Monitoring the Health-Related Sustainable Development Goals: Lessons Learned and Recommendations for Improved Measurement. *The Lancet* 395: 240–46. [CrossRef]

Barbazza, Erica, and Juan E. Tello. 2014. A Review of Health Governance: Definitions, Dimensions and Tools to Govern. *Health Policy* 116: 1–11. [CrossRef]

Beaglehole, Robert, and Ruth Bonita. 2010. What Is Global Health? *Global Health Action* 3. [CrossRef]

Bennett, Belinda, I. Glenn Cohen, Sara E. Davies, Lawrence O. Gostin, Peter S. Hill, Aditi Mankad, and Alexandra L. Phelan. 2018. Future-Proofing Global Health: Governance of Priorities. *Global Public Health* 13: 519–27. [CrossRef] [PubMed]

Berner-Rodoreda, Astrid, Eva Annette Rehfuess, Kerstin Klipstein-Grobusch, Frank Cobelens, Mario Raviglione, Antoine Flahaut, Núria Casamitjana, Günter Fröschl, Jolene Skordis-Worral, Ibrahim Abubakar, and et al. 2019. Where Is the 'Global' in the European Union's Health Research and Innovation Agenda? *BMJ Global Health* 4: e001559. [CrossRef] [PubMed]

Bettcher, Douglas, and Kelley Lee. 2002. Globalisation and Public Health. *Journal of Epidemiology and Community Health* 56: 8–17. [CrossRef] [PubMed]

Blue, Laura, and Thomas J. Espenshade. 2011. Population Momentum Across the Demographic Transition. *Population and Development Review* 37: 721–47. [CrossRef]

Boutayeb, Abdesslam. 2006. The Double Burden of Communicable and Non-Communicable Diseases in Developing Countries. *Transactions of the Royal Society of Tropical Medicine and Hygiene* 100: 191–99. [CrossRef]

Bozorgmehr, Kayvan. 2010. Rethinking the 'Global' in Global Health: A Dialectic Approach. *Globalization and Health* 6: 1–19. [CrossRef] [PubMed]

Bozorgmehr, Kayvan, and Oliver Razum. 2016. Refugees in Germany—Untenable Restrictions to Health Care. *The Lancet* 388: 2351–52. [CrossRef]

Bozorgmehr, Kayvan, and Katharina Wahedi. 2017. Reframing Solidarity in Europe: Frontex, Frontiers, and the Fallacy of Refugee Quota. *The Lancet Public Health* 2: e10–e11. [CrossRef]

Burstein, Roy, Nathaniel J. Henry, Michael L. Collison, Laurie B. Marczak, Amber Sligar, Stefanie Watson, Neal Marquez, Mahdieh Abbasalizad-Farhangi, Masoumeh Abbasi, Foad Abd-Allah, and et al. 2019. Mapping 123 Million Neonatal, Infant and Child Deaths Between 2000 and 2017. *Nature* 574: 353–58. [CrossRef]

Byskov, Jens, Stephen Maluka, Bruno Marchal, Elizabeth H. Shayo, Astrid Blystad, Salome Bukachi, Joseph M. Zulu, Charles Michelo, Anna-Karin Hurtig, and Paul Bloch. 2019. A Systems Perspective on the Importance of Global Health Strategy Developments for Accomplishing Today's Sustainable Development Goals. *Health Policy and Planning.* 34: 635–45. [CrossRef]

Campbell-Lendrum, Diarmid, and Carlos Corvalán. 2007. Climate Change and Developing-Country Cities: Implications for Environmental Health and Equity. *Journal of Urban Health* 84 S3: 109–17. [CrossRef]

Carlson, Valeria, Marita J. Chilton, Liza C. Corso, and Leslie M. Beitsch. 2015. Defining the Functions of Public Health Governance. *American Journal of Public Health* 105 S2: S159–S166. [CrossRef]

Chan, Margaret. 2013. Investing in Health: Progress but Hard Choices Remain. *The Lancet* 382: e34–e35. [CrossRef]

Chang, Angela Y., Krycia Cowling, Angela E. Micah, Abigail Chapin, Catherine S. Chen, Gloria Ikilezi, Nafis Sadat, Golsum Tsakalos, Junjie Wu, Theodore Younker, and et al. 2019. Past, Present, and Future of Global Health Financing: A Review of Development Assistance, Government, Out-of-Pocket, and Other Private Spending on Health for 195 Countries, 1995–2050. *The Lancet* 393: 2233–60. [CrossRef]

Clift, Charles, and John-Arne Røttingen. 2018. New Approaches to WHO Financing: The Key to Better Health. *BMJ* 361: k2218. [CrossRef] [PubMed]

Dockweiler, Christoph, and Florian Fischer, eds. 2019. *ePublic Health*. Bern: Hogrefe.

Donkin, Angela, Peter Goldblatt, Jessica Allen, Vivienne Nathanson, and Michael Marmot. 2018. Global Action on the Social Determinants of Health. *BMJ Global Health* 3: S1. [CrossRef]

Dye, Christopher. 2014. After 2015: Infectious Diseases in a New Era of Health and Development. *Philosophical Transactions of the Royal Society of London. Series B Biological Sciences* 369: 20130426. [CrossRef]

European Commission. 2010. *Communication from the Commission to the Council, the European Parliament, the European Economic and Social Commitee and the Committee of the Regions. The EU Role in Global Health*. Brussels: European Commission.

Flahault, Antoine, Antoine Geissbuhler, Idris Guessous, Philippe Guérin, Isabelle Bolon, Marcel Salathé, and Gérard Escher. 2017. Precision Global Health in the Digital Age. *Swiss Medical Weekly* 147: w14423. [CrossRef]

Friel, Sharon, Michael Marmot, Anthony J. McMichael, Tord Kjellstrom, and Denny Vågerö. 2008. Global Health Equity and Climate Stabilisation: A Common Agenda. *The Lancet* 372: 1677–83. [CrossRef]

Gautier, Lara, Isidore Sieleunou, and Albino Kalolo. 2018. Deconstructing the Notion of Global Health Research Partnerships Across Northern and African Contexts. *BMC Medical Ethics* 19: 13–20. [CrossRef] [PubMed]

GAVI. 2019. *Annual Progress Report 2018*. Geneva: GAVI.

Glatman-Freedman, Aharona, and Katherine Nichols. 2012. The Effect of Social Determinants on Immunization Programs. *Human Vaccines & Immunotherapeutics* 8: 293–301. [CrossRef]

Gostin, Lawrence O., Benjamin M. Meier, Rebekah Thomas, Veronica Magar, and Tedros A. Ghebreyesus. 2018. 70 Years of Human Rights in Global Health: Drawing on a Contentious Past to Secure a Hopeful Future. *The Lancet* 392: 2731–35. [CrossRef]

Greenwood, Brian. 2014. The Contribution of Vaccination to Global Health: Past, Present and Future. *Philosophical Transactions of the Royal Society of London. Series B, Biological Sciences* 369: 20130433. [CrossRef]

Hay, Simon I., Puja C. Rao, Christiane Dolecek, Nicholas P. J. Day, Andy Stergachis, Alan D. Lopez, and Christopher J. L. Murray. 2018. Measuring and Mapping the Global Burden of Antimicrobial Resistance. *BMC Medicine* 16: 78. [CrossRef]

Honigsbaum, Mark. 2017. Between Securitisation and Neglect: Managing Ebola at the Borders of Global Health. *Medical History* 61: 270–94. [CrossRef]

Huynen, Maud M. T. E., Pim Martens, and Henk B. M. Hilderink. 2005. The Health Impacts of Globalization: A Conceptual Framework. *Globalization and Health* 1: 14. [CrossRef] [PubMed]

IOM. 2018. *World Migration Report 2018*. Geneva: International Organization for Migration.

Jamison, Dean T., Lawrence H. Summers, George Alleyne, Kenneth J. Arrow, Seth Berkley, Agnes Binagwaho, Flavia Bustreo, David Evans, Richard G. A. Feachem, Julio Frenk, and et al. 2013. Global Health 2035: A World Converging Within a Generation. *The Lancet* 382: 1898–955. [CrossRef]

Juárez, Sol Pía, Helena Honkaniemi, Andrea C. Dunlavy, Robert W. Aldridge, Mauricio L. Barreto, Srinivasa Vittal Katikireddi, and Mikael Rostila. 2019. Effects of Non-Health-Targeted Policies on Migrant Health: A Systematic Review and Meta-Analysis. *The Lancet Global Health* 7: e420–35. [CrossRef]

Kellehear, Allan. 2017. Public Health Approaches to Dying, Death, and Loss. In *International Encyclopedia of Public Health*, 2nd ed. Edited by Stella R. Quah and William C. Cockerham. Oxford: Academic Press, pp. 184–89. Available online: http://www.sciencedirect.com/science/article/pii/B9780128036785003659 (accessed on 17 October 2019).

Keusch, Gerald T., Wen L. Kilama, Suerie Moon, Nicole A. Szlezák, and Catherine M. Michaud. 2010. The Global Health System: Linking Knowledge with Action–Learning from Malaria. *PLoS Medicine* 7: e1000179. [CrossRef]

Kickbusch, Ilona. 2014. Governance for Health, Wellbeing and Sustainability—What Is at Stake. *Global Health Promotion* 21 S1: 83. [CrossRef]

Kickbusch, Ilona. 2016. Global Health Governance Challenges 2016—Are We Ready? *International Journal of Health Policy and Management* 5: 349–53. [CrossRef]

Kirk, Dudley. 1996. Demographic Transition Theory. *Population Studies* 50: 361–87. [CrossRef]

Koplan, Jeffrey P., T. Christopher Bond, Michael H. Merson, K. Srinath Reddy, Mario Henry Rodriguez, Nelson K. Sewankambo, and Judith N. Wasserheit. 2009. Towards a Common Definition of Global Health. *The Lancet* 373: 1993–95. [CrossRef]

Laaser, Ulrich, and Leon Epstein. 2010. Threats to Global Health and Opportunities for Change: A New Global Health. *Public Health Reviews* 32: 54–89. [CrossRef]

Labrique, Alain, Lavanya Vasudevan, Garrett Mehl, Ellen Rosskam, and Adnan A. Hyder. 2018. Digital Health and Health Systems of the Future. *Global Health: Science and Practice* 6 S1: S1–S4. [CrossRef]

Laxminarayan, Ramanan, Adriano Duse, Chand Wattal, Anita K. M. Zaidi, Heiman F. L. Wertheim, Nithima Sumpradit, Erika Vlieghe, Gabriel Levy Hara, Ian M. Gould, Herman Goossens, and et al. 2013. Antibiotic Resistance—The Need for Global Solutions. *The Lancet Infectious Diseases* 13: 1057–98. [CrossRef]

Lee, Kelley. 2003. *Globalization and Health: An Introduction*. Global Issues Series; Houndmills, Basingstoke, Hampshire, and New York: Palgrave Macmillan.

Lee, Kelley. 2004. Globalisation: What Is It and How Does It Affect Health? *Medical Journal of Australia* 180: 156–58. [CrossRef]

Liang, Lu, and Peng Gong. 2017. Climate Change and Human Infectious Diseases: A Synthesis of Research Findings from Global and Spatio-Temporal Perspectives. *Environment International* 103: 99–108. [CrossRef]

Lozano, Rafael, Nancy Fullman, Degu Abate, Solomon M. Abay, Cristiana Abbafati, Nooshin Abbasi, Hedayat Abbastabar, Foad Abd-Allah, J. Abdela, A. Abdelalim, and et al. 2018. Measuring Progress from 1990 to 2017 and Projecting Attainment to 2030 of the Health-Related Sustainable Development Goals for 195 Countries and Territories: A Systematic Analysis for the Global Burden of Disease Study 2017. *The Lancet* 392: 2091–138. [CrossRef]

Machalaba, Catherine, Cristina Romanelli, Peter Stoett, Sarah E. Baum, Timothy A. Bouley, Peter Daszak, and William B. Karesh. 2015. Climate Change and Health: Transcending Silos to Find Solutions. *Annals of Global Health* 81: 445–58. [CrossRef]

Marmot, Michael, and Ruth Bell. 2019. Social Determinants and Non-Communicable Diseases: Time for Integrated Action. *BMJ* 364: l251. [CrossRef]

Martin, Keith, Zoë Mullan, and Richard Horton. 2014. An Evolution in Global Health. *The Lancet Global Health* 2: S1–S2. [CrossRef]

Mason, Paul H., Ian Kerridge, and Wendy Lipworth. 2017. The Global in Global Health Is Not a Given. *The American Journal of Tropical Medicine and Hygiene* 96: 767–69. [CrossRef]

McBride, Bronwyn, Sarah Hawkes, and Kent Buse. 2019. Soft Power and Global Health: The Sustainable Development Goals (SDGs) Era Health Agendas of the G7, G20 and BRICS. *BMC Public Health* 19: 1–14. [CrossRef]

McCoy, David, Sudeep Chand, and Devi Sridhar. 2009. Global Health Funding: How Much, Where It Comes from and Where It Goes. *Health Policy and Planning* 24: 407–17. [CrossRef]

McNeil, Candice J., and Avinash K. Shetty. 2017. Zika Virus: A Serious Global Health Threat. *Journal of Tropical Pediatrics* 63: 242–48. [CrossRef] [PubMed]

Means, Arianna R., David E. Phillips, Grégoire Lurton, Anne Njoroge, Sabine M. Furere, Rong Liu, Wisal M. Hassan, Xiaochen Dai, Orvalho Auguto, Peter Cherutich, and et al. 2016. The role of implementation science training in global health: From the perspective of graduates of the field's first dedicated doctoral program. *Global Health Action* 9: 1. [CrossRef]

Meier, Benjamin Mason, Dabney P. Evans, Matthew M. Kavanagh, Jessica M. Keralis, and Gabriel Armas-Cardona. 2018. Human Rights in Public Health: Deepening Engagement at a Critical Time. *Health and Human Rights* 20: 85–91. [PubMed]

Moallemi, Enayat A., Shirin Malekpour, Michalis Hadjikakou, Rob Raven, Katrina Szetey, Mehran Mahdavi Moghadam, Reihaneh Bandari, Rebecca Lester, and Brett A. Bryan. 2019. Local Agenda 2030 for Sustainable Development. *The Lancet Planetary Health* 3: e240–e241. [CrossRef]

National Academies of Sciences, Engineering and Medicine. 1992. *Global Environmental Change*. Washington: National Academies Press.

Nature. 2020. Editorial: Getting out of the World Health Organization might not be as easy as Trump thinks. *Nature* 582: 459. [CrossRef]

Oke, Nicole. 2009. Globalizing Time and Space: Temporal and Spatial Considerations in Discourses of Globalization. *International Political Sociology* 3: 310–26. [CrossRef]

Omran, Abdel R. 1971. The Epidemiologic Transition. A Theory of the Epidemiology of Population Change. *The Milbank Memorial Fund Quarterly* 49: 509–38. [CrossRef]

Patz, Jonathan A., Howard Frumkin, Tracey Holloway, Daniel J. Vimont, and Andrew Haines. 2014. Climate Change: Challenges and Opportunities for Global Health. *JAMA* 312: 1565–80. [CrossRef] [PubMed]

Raguin, Gilles, and Pierre-Marie Girard. 2018. Toward a Global Health Approach: Lessons from the HIV and Ebola Epidemics. *Globalization and Health* 14: 114. [CrossRef]

Reddy, Srikanth K., Sumaira Mazhar, and Raphael Lencucha. 2018. The Financial Sustainability of the World Health Organization and the Political Economy of Global Health Governance: A Review of Funding Proposals. *Globalization and Health* 14: 11. [CrossRef] [PubMed]

Reidpath, Daniel D., and Pascale Allotey. 2007. Measuring Global Health Inequity. *International Journal for Equity in Health* 6: 1–7. [CrossRef]

Ridde, Valéry. 2016. Need for More and Better Implementation Science in Global Health. *BMJ Global Health* 1: e000115. [CrossRef]

Ridde, Valéry, Dennis Pérez, and Emilie Robert. 2020. Using Implementation Science Theories and Frameworks in Global Health. *BMJ Global Health* 5: e002269. [CrossRef]

Roe, Dilys. 2019. Biodiversity Loss—More Than an Environmental Emergency. *The Lancet Planetary Health* 3: e287–e289. [CrossRef]

Rowthorn, Virginia. 2015. Global/Local: What Does It Mean for Global Health Educators and How Do We Do It? *Annals of Global Health* 81: 593–601. [CrossRef]

Rudan, Igor, and Devi Sridhar. 2016. Structure, Function and Five Basic Needs of the Global Health Research System. *Journal of Global Health* 6: 10505. [CrossRef]

Sachs, Jeffrey, Guido Schmidt-Traub, C. Kroll, Guillaume Lafortune, and Grayson Fuller. 2019. *Sustainable Development Report 2019: Transformations to Achieve the Sustainable Development Goals*. New York: Bertelsmann Stiftung and Sustainable Development Solutions Network.

Schrecker, Ted, Ronald Labonté, and Roberto de Vogli. 2008. Globalisation and Health: The Need for a Global Vision. *The Lancet* 372: 1670–76. [CrossRef]

Shadmi, Efrat, Yingyao Chen, Inês Dourado, Inbal Faran-Perach, John Furler, Peter Hangoma, Piya Hanvoravongchai, Claudia Obando, Varduhi Petrosyan, Krishna D. Rao, and et al. 2020. Health Equity and COVID-19: Global Perspectives. *International Journal for Equity in Health* 19: 104. [CrossRef]

Sinclair, Julie R. 2019. Importance of a One Health Approach in Advancing Global Health Security and the Sustainable Development Goals. *International Office of Epizootics* 38: 145–54. [CrossRef]

Stringhini, Silvia, and Pascal Bovet. 2017. Socioeconomic Status and Risk Factors for Non-Communicable Diseases in Low-Income and Lower-Middle-Income Countries. *The Lancet Global Health* 5: e230–e231. [CrossRef]

The Lancet. 2012. Global Health in 2012: Development to Sustainability. *The Lancet* 379: 193. [CrossRef]

Tilford, Sylvia. 2018. From Alma Ata 1978 to Almaty 2018. *International Journal of Health Promotion and Education* 56: 262–64. [CrossRef]

UN. 2015a. *The Millennium Development Goals Report 2015*. New York: United Nations.

UN. 2015b. *Transforming Our World: The 2030 Agenda for Sustainable Development*. New York: United Nations.

UN. 2016. Goal 3—Targets and Indicators: Sustainable Development Knowledge Platform. Available online: https://sustainabledevelopment.un.org/sdg3#targets (accessed on 26 October 2019).

UN. 2019a. Progress Towards the Sustainable Development Goals. Available online: https://sustainabledevelopment.un.org/sdg3 (accessed on 29 September 2019).

UN. 2019b. SDG Acceleration Actions: Sustainable Development Knowledge Platform. Available online: https://sustainabledevelopment.un.org/sdgactions (accessed on 26 October 2019).

Van Belle, Sara, Remco van de Pas, and Bruno Marchal. 2017. Towards an agenda for implementation science in global health: There is nothing more practical than good (social science) theories. *BMJ Global Health* 2: e000181. [CrossRef] [PubMed]

Walraven, Gijs. 2019. The 2018 Astana Declaration on Primary Health Care, Is It Useful? *Journal of Global Health* 9: 10313. [CrossRef] [PubMed]

Wanless, Derek. 2004. *Securing Good Health for the Whole Population: Final Report*; With the Assistance of Natasha Jones, Robert Anderson, Frank Bowley, Sunjai Gupta, John Hall, Ruth Lawson, Bhash Naidoo, Mary Jane Roberts, Harpa Sabharwal, Jeremy Skinner and et al. London: HM Treasury. Available online: https://www.southampton.gov.uk/moderngov/documents/s19272/prevention-appx%201%20wanless%20summary.pdf (accessed on 28 October 2019).

Watkins, Richard R., and Robert A. Bonomo. 2016. Overview: Global and Local Impact of Antibiotic Resistance. *Infectious Disease Clinics of North America* 30: 313–22. [CrossRef]

Watts, Nick, Markus Amann, Nigel Arnell, Sonja Ayeb-Karlsson, Kristine Belesova, Maxwell Boykoff, Peter Byass, Wenjia Cai, Diarmid Campbell-Lendrum, Stuart Capstick, and et al. 2019. The 2019 Report of the Lancet Countdown on Health and Climate Change: Ensuring That the Health of a Child Born Today Is Not Defined by a Changing Climate. *The Lancet* 394: 1836–78. [CrossRef]

West-Oram, Peter G. N., and Alena Buyx. 2017. Global Health Solidarity. *Public Health Ethics* 10: 212–24. [CrossRef]

WHO. 1946. *Constitution of the World Health Organization: Preamble*. Geneva: World Health Organization.

WHO. 1978. *Declaration of Alma-Ata*. Geneva: World Health Organization.

WHO. 2013. *Global Vaccine Action Plan 2011–2020*. Geneva: World Health Organization.

WHO. 2014. *Health in All Policies: Framework for Country Action*. Geneva: World Health Organization.

WHO. 2018a. *Declaration of Astana*. Geneva: World Health Organization.

WHO. 2018b. *World Health Statistics 2018: Monitoring Health for the SDGs*. Geneva: World Health Organization.

WHO. 2019a. The Role of WHO in Public Health. Available online: https://www.who.int/about/role/en/ (accessed on 28 October 2019).

WHO. 2019b. *World Health Statistics 2019: Monitoring Health for the SDGs*. Geneva: World Health Organization.

WHO Europe. 2019. *Policy Coherence as a Driver of Health Equity*. Copenhagen: World Health Organization—Regional Office for Europe.

Wickramage, Kolitha, Jo Vearey, Anthony B. Zwi, Courtland Robinson, and Michael Knipper. 2018. Migration and Health: A Global Public Health Research Priority. *BMC Public Health* 18: 1–9. [CrossRef]

Yang, Alice, Paul E. Farmer, and Anita M. McGahan. 2010. 'Sustainability' in Global Health. *Global Public Health* 5: 129–35. [CrossRef] [PubMed]

Zaidi, Batool, and S. Philip Morgan. 2017. The Second Demographic Transition Theory: A Review and Appraisal. *Annual Review of Sociology* 43: 473–92. [CrossRef] [PubMed]

On the Relationship between Health Sectors' Digitalization and Sustainable Health Goals: A Cyber Security Perspective

Stefan Sütterlin, Benjamin J. Knox, Kaie Maennel, Matthew Canham and Ricardo G. Lugo

1. Towards Sustainable Digitalized Healthcare

Digitalization in the health sector is, as in all societal domains, motivated by a range of anticipated positive consequences, such as increased effectiveness of prevention, treatment and follow-up, a generally improved resource efficiency and improved health care availability. This chapter will discuss how the ambition of achieving sustainable health goals may be affected by measures of digitalization. This will be done by covering digitalization from a cyber security perspective and how new potential threats to privacy may influence the public's trust in their health care system, thereby affecting the envisaged goals of sustainable health care performance. It will also discuss further how digitalization in the healthcare sector unleashes an enormous potential in terms of cost-effectiveness, decentralization and the availability of specialist services and expertise, which risks and countermeasures these changes entail, and how they are currently dealt with and the role of cyber resilience in ensuring rapid digitalization does not come at the cost of essential trust mechanisms that are quid pro quo in the health sector.

Transformation to a digitalized healthcare system must be governed and framed by a range of measures in various societal domains. The World Health Organization's (WHO) report on the status of eHealth in the European region (WHO 2016) states that its member states "acknowledge and understand the role of e-Health in contributing to the achievement of universal health coverage and have a clear recognition of the need for national policies strategies and governance to ensure the progress and long-term sustainability of investments. However, leveraging eHealth as a national strategic asset demands a more coordinated approach [...]" (WHO 2016, p. xi). The WHO acknowledges the fact that a majority of its member states developed or are developing "national strategies or policies for eHealth, universal health coverage or national health information systems, and ensuring sustainable funding for their implementation." (WHO 2016, p. xi). The report requests the member states to develop an "inclusive and intersectorial approach to the development of national eHealth strategies" (p. xii). These international efforts to adapt to the consequences of technological development provide an example for the wide ranging consequences that the digitalization of health systems entails for legal security measures relating to a state's power over its citizens, institutional development, policy and practice, education, and cyber security. We argue that cyber security, in turn, is not just one of many subjects to consider, but instead it pertains to and penetrates all other affected domains, such as the aforementioned organizational structures, policymaking, and education of non-technical healthcare professionals, as well as legal structures.

The healthcare system's efficiency and effectiveness depends on policies and procedures such as the rapid and precise exchange of health-related information between its different acteurs. In recent years, more national healthcare systems approached sustainable health benefits for their populations by digitizing their services and administrative procedures, enabling new innovative models of health care delivery and efficient data sharing amongst stakeholders. Digitalization on a system level led to increasingly centralized databases offering synergies for new research possibilities, and the opportunity to improve individualised, time- and cost-efficient healthcare covering hitherto underserviced areas, such as disadvantaged parts of society or structurally weak geographical spheres. The manifold effects of digitalized healthcare result from a large variety of technological innovations on individual, institutional or system level.

The concept of e-Health combines aspects of areas such as healthcare business and administration, public health, and medical informatics. This broad term covers aspects of healthcare delivery, data administration and relies, to a large extent, on the availability of web-based patient care via communication platforms. e-Health is both a technical tool of practical healthcare delivery as well as a governance instrument. The term is also used as a more general description of a larger variety of digitalized healthcare. On the contrary, the term m-health ("mobile health") is a particular aspect of e-health describing medical and public health practice via mobile (wireless) electronic devices such as smartphones or specific monitoring devices such as EKG monitors or glucometers. Telehealth is a rather broad term referring to remote clinical services, as well as non-clinical activities such as training and medical education. The narrower term telemedicine is the aspect of telehealth dealing with remote clinical services. These various manifestations of digitalized healthcare contribute to the gradual achievement of sustainability goals in a number of ways. Within the last ten to fifteen years, electronic health records have become a universal feature of digitalized healthcare. This increased availability of digitalized information and the overall improved information management by healthcare professionals accelerated decision-making time, provided an information basis for better outcomes of clinical decisions, facilitated interdisciplinary cooperation, reduced the fragmentation of information between healthcare providers, and reduced duplicate tests and other treatment risks or side effects (Atasoy et al. 2019).

In developing countries, digitalization and thus this universally available information has increasingly been used not only to process patient-related information, but also to make medical services and the competence of rare and remote health care providers accessible to rural and secluded regions with lacking or insufficient medical infrastructure (Zhang and Zaman 2019) by means of eHealth. While eHealth has been mostly seen in terms of reducing inequality to access of healthcare, a discussion on unequal access to these benefits emerges. As mHealth contributes to an increased autonomy and strengthening of the patient role as administrator and gatekeeper over own health-related data, this increases the demands towards patients considerably. The access to "patient-facing" mHealth and a consequently more responsible and autonomous patient role are raising concerns related to digital inequality and cognitive skills. Unequal access to the benefits of digitalization for deprived societal groups could be particularly pronounced in countries with rather limited universal healthcare, such as the USA or developing countries.

2. Digitalization and Increased Vulnerability

Highly digitized healthcare systems relying on electronic health records are a vital part of a nation's critical infrastructure. Critical infrastructure describes the "physical and cyber systems and assets that are so vital […] that their capacity or destruction would have a debilitating impact on […] physical or economic security or public health or safety" (Department of Homeland Security 2019).

The model of an increasingly patient-centered healthcare emphasizing the individual's needs, rights, autonomy, preferences and values guiding all medical decisions is increasingly promoted and spread as a development unfolding parallel to digital change. Healthcare systems involve patients to an increased extent in ambulatory data collection, provide more insight into patient documentation, secure the patient's rights to own the data and become more inclusive when choosing amongst treatment options. As an effect of the increased cooperation between patient and numerous other acteurs in the healthcare system, the increased amount of patient-centered sensitive data is bundled, stored and processed in centralized or highly interconnected digital infrastructures.

The amount and availability of patient-related data are tempting high-value assets for criminal undertakings. The extent to which the healthcare sector has become a primary target of cyber crime is reflected in the number of cyber attacks on hospitals in recent years. More than 2000 breaches involving a total of 176 million patient records were reported between 2010 and 2017, with individual breaches ranging from 500 to nearly 79 million patient records. The total number of breaches increased from 199 in 2010 to 344 in 2017 (e.g., McCoy and Perlis 2018). One of the first occasions when the international general public became aware of cyber attacks on the health care sector was the 2017 WannaCry ransomware attack infecting an estimated 230,000 windows systems in 150 countries (Chen and Bridges 2017). The unprecedented scale of this cyber attack brought the relatively neglected field of cyber security in the healthcare sector onto the agenda of media, healthcare providers and policy makers.

The many years of negligence towards cyber security made many healthcare providers particularly easy targets, especially when compared to highly protected societal domains such as the military, law enforcement, and innovation-driven economy safeguarding their intellectual property or high-value tactical or strategic decisions. The Norwegian data breach of 2018 is a particular—but not exclusive—example of a targeted attack with the criminal intent to obtain sensitive patient data exploiting the lack of awareness and preparedness. The targeted attack resulted in the extraction of 2.9 million datasets, including patient records, resembling more than half of Norway's population (Hughes 2018; Johansson 2018). The attack is believed to have been carried out by highly sophisticated attackers explicitly targeting patient records stored by the health authorities in the South-East region of Norway. Following further investigations, it was assumed that the attackers were focused on the health service's relationship with Norway's armed forces and potential information regarding a major NATO exercise organized in Norway later in the year. The cyber attack seems to have benefitted from lax security standards and outdated operating systems no longer supported by security updates. According to the investigations of a national newspaper, as late as in 2017, about 1200 of the Health-South-East's computers were still running on Windows XP, about three years after Microsoft ceased supporting the popular operating system (Irwin 2018; VG 2017).

While these events demonstrate clearly the need for heightened cyber security awareness and preparedness, the viciousness of criminal hackers—be they rogue criminals with economic interests, financed by nation-states, competitors or others—data breaches do not necessarily presuppose criminal intent. Human failure is known to be the major vulnerability of all technical systems with which humans interact ("socio-technical systems"). Human failure provides numerous "attack vectors", i.e., opportunities to exploit a person's individual vulnerability in order to access a targeted network. However, while processes, people and technology need to be designed, trained, and maintained in a way that reduces the risk of human failure enabling third parties with malicious intent to get access to sensitive data, the loss of sensitive patient data does not necessarily require malicious intent. Amongst some of the most prominent—known—health data breaches made public in recent years include the breach of data on 150,000 patients by the UK's National Health Service being shared without their permission in 2018. In fact, the patients whose data were made public actively opted out of any use of their data beyond own treatment. The breach was later explained as a "coding error", i.e., an internal human failure without any criminal intent (Canham et al. 2020). The National Health Service (NHS) data breach of 2018 indicates how cyber security breaches and efforts to improve information security management routines are of crucial relevance for the patients' trust in the healthcare system, even in the absence of any criminal attempt, as it undermines the public trust in the institutions' very ability to maintain secure information processing as a part of their daily routines, and in the absence of acute threats. The case of the NHS data breach demonstrated impressively how a digitalized data administration potentiates the detrimental impact of singular human failure.

The motivation for targeting the health sector is comparable to other large societal sectors such as public authorities, governmental or non-governmental organizations or industry. With cybercrime dominating, hacktivism and cyber espionage are the major motivations behind cyber attacks in the healthcare sector (Passeri 2017). The often-reported observation that human factors seem to play a particular role in cyber breaches targeting the healthcare sector, has been associated with the particular lack of security awareness, training, and security-related attitudes (Kim 2017). As a result, health and hospital trusts are frequently considered "low hanging fruits" by cyber criminals aiming for financial profits on the black market for patient data. In addition, and in the comparably smaller but potentially more devastating case of nation-actors, the healthcare sector provides a relatively easy entrance for the acquisition of sensitive information required to paralyze a potential enemy's critical healthcare infrastructure in a conflict situation.

Cyber security rests upon three pillars; confidentiality, availability, and integrity (Conrad et al. 2012; Canham et al. 2018). Cyber-attacks focused on the confidentiality pillar expose private information (such as credit card or health information) to unauthorized persons. The American Medical Collection Agency (AMCA) data breach in which 12 million patient records were exposed is an example of an attack using the confidentiality pillar. The integrity pillar represents the prevention of unauthorized modification of a system or data. Attacks against the integrity pillar modify data in such a way that it becomes untrustworthy, and the attack may not be discovered even after the damage has been done. In a healthcare context, cyber threat actors who attack the integrity pillar might change a patient's dosage instructions and create a potentially unsafe situation for that patient. The Stuxnet worm is often cited as an example of an attack on information integrity that caused a large number of Iranian uranium centrifuges to malfunction. This attack employed malware specifically designed to provide false performance readings to operators that left them unaware of an impending problem (Chien et al. 2012). Security researchers have conducted proof-of-concept attacks against the Integrity Pillar of neuro-prosthetic devices and were able to inject commands that disrupted the device's normal functionality (Canham and Sawyer 2019; Cusack et al. 2017). The availability pillar focuses on information assurance; in other words, ensuring that users are able to access the data that they are authorized to access. Examples of attacks against the availability pillar include the multitude of ransomware attacks against hospitals and healthcare facilities over the past few years.

The examples of major data breaches provided above are only a very small part of the overall picture and do by no means cover a substantial part of known cyber threats. Commercial cyber security providers publish regular lists of conducted cyber attacks that have become known. At the time point of detection, cyber attacks have usually already happened, and a considerable number of attacks and conducted data extractions may remain undetected or unreported. The majority of attacks directly target individual hospitals as key centers of gravity, with direct access to patients' sensitive data, while being known to be lagging far behind other institutions of comparable size in terms of their cyber security standards. Where a few hospitals with low resources for cyber security threaten the entire infrastructure due to their interconnectedness, the efforts to become a less attractive goal for cyber criminals needs to step up considerably. These efforts need to go beyond current regulations which largely focus on questions related to data privacy, but not data security (Jalali and Kaiser 2018). A surge in related analyses and systematic research on risk factors and potential countermeasures in recent years indicates an increasing awareness amongst the stakeholders, which may lead to increased future investment in technology, processes and personnel (Coventry and Branley 2018; Kruse et al. 2017; Martin et al. 2017).

3. Improving Sustainable Healthcare and Ensuring Cybersecurity: The Estonian Case

As digitalization as a means to improve national healthcare systems' effectiveness and efficiency comes with increased data exchange accompanied by an increased potential of vulnerabilities towards security threats, governments work towards finding a balance between technically feasible solutions maximizing the outcome benefits, whilst minimizing the threat potential with cyber security measures. Cyber resilient healthcare systems contribute to and maintain public acceptance for further transformation and sustainable development initiatives, and bolster public perception of information security. When discussing these future trends, the digitalization of the Estonian healthcare system may serve as a blueprint for future developments in other countries, too. The Baltic country with approximately 1.2 million inhabitants is considered a world leader in both e-government as well as cyber security. Estonia is regarded to be one of the most digitized countries of the world and has firsthand experience of being subject to hostile cyber attacks by a foreign state power. In 2007, a Russian cyber attack brought the Estonian public authorities, organizations, media and with it, the usual public life, to an abrupt hold. The attacks were interpreted as an expression of Russia's disagreement with the relocation of a Sovjet-era bronze soldier statue in the capital city Tallinn. In the aftermath of these attacks and reinforced by a continuously perceived threat by Russian interference, the country publicly and privately invested heavily in cyber security infrastructure, processes and competences. Both the level of digitalization of the country's healthcare services and the technological advancements in cyber security make Estonia an interesting case study and "laboratory" providing possible insights into future developments on an international scale. Trends and consequences in Estonia are likely to be—in similar or partial form—also introduced in other countries.

Estonia's healthcare system's patients and doctors, hospitals and the government, have to rely on e-services for health. Each person in Estonia that has visited a doctor at least once has an online e-Health record. The core of the Estonian e-Health System is the digital health record that functions as a centralized, national database and retrieves data as necessary from various providers, who may be using different systems, and presents it in a standard format via the e-Patient portal (e-estonia.com). These e-Health records are, among other purposes, used for prescriptions and in emergency situations. E-ambulances are available in emergency situations to detect and locate an emergency call within 30 seconds. On board the e-ambulance, a doctor can use a patient's ID code to read time-critical information, such as blood type, allergies, or recent treatments. To balance these obvious advantages in availability, efficiency and cost-effectiveness on the societal level, data privacy for the individual is secured by various means. Sensitive patient information is only accessible by authorised individuals identified via national electronic ID cards. Highly secure state-of-the art data transmission encryption techniques (Keyless Signature Infrastructure (KSI) Blockchain technology) are used for the system to ensure data integrity and mitigate internal threats to the data. Data transport and security layers are provided for by the government's data transmission software ("X-Road"), ensuring that only digitally signed and encrypted data are exchanged (Priisalu and Ottis 2017). The electronic databases are organised in a decentralized manner ensuring that cyber attacks or leaks cannot compromise the overall system. There are no so-called super-administrators with unrestricted access to patients' health records (Health and Welfare Information Systems Centre (TEHIK) (2019)). By default, medical specialists can access data using their unique individual identifier, but any patient can choose to deny access to any case-related data, to any, or all care providers; including one's own general practitioner (Priisalu and Ottis 2017). Every patient thus keeps full autonomic control over the stored data, provides permission and keeps the full overview over which individual person has accessed their personal information. Each data view leaves an identifiable trace and record.

The privacy and processing of personal data processed in the Estonian healthcare system and other public databases is regulated by the country's Personal Data Protection Act and the European Union's General Data Protection Regulation (GDPR). The governing principle is to provide an open and transparent attitude as a prerequisite for a robust trusted relationship between the citizens and the state. By investing in a person's confidence in the government's ability to keep their data secure and guarantee confidentiality, integrity and availability. While there are no direct survey data describing the public and individual trust into the Estonian eHealth services available, the high rate of access and usage since the system's introduction in December 2008, the constantly high ratings of perceived quality of healthcare in public surveys indicate a level of acceptance (Lai et al. 2013). The Estonian example combines state-of-the art e-government services with newest cyber security technology, maximizing the potential of "security by design", i.e., providing technology with least possible vulnerabilities and robustness towards human failure. To ensure public acceptance and trust, the patient is given full autonomy and transparency over the use of his/her personal information.

There are currently no known comprehensive data breaches following attacks on the Estonian healthcare system. While security-by-design solutions are a powerful and necessary means to provide cyber resilience, the human factor remains a considerable risk factor, particularly in less advanced partially digitalized systems and healthcare systems with lower degrees of centralization, as typical in larger countries. Human behaviour thus remains a constant threat of cyber security that is subject to intensifying research efforts by behavioural and interdisciplinary scientists.

4. The Human Factor in the Healthcare System's Cyber Security

Research on the human factor in cyber security acknowledges that technology does not exist in isolation, but that interpretations, conclusions and decisions are made by individuals or groups of humans with the "inbuilt guarantee" to commit a whole range of human failures if given the opportunity. Thus, even the best designed system for processing and storing sensitive data in a healthcare system faces human users and thus human failures, which potentially compromise data security at least on an individual, if not even systemic level, in unforeseen ways. These human failures—such as the coding error enabling the 2018 NHS data breach—occur on individual levels following erroneous conclusions and decision-making processes, insufficient or biased information as a decision-making foundation following inappropriate or inaccurate communication between individuals, teams, institutions or authorities. Given the relatively recent awakening concerning awareness of cyber vulnerabilities in the healthcare sectors, we argue that the human factor based on training, education and compliance, is of particular importance. In areas such as aviation, acute medical care, and many safety- and security-critical sectors, the devastating effects of human failure, for example by miscommunications, are well documented and acknowledged. In the area of cyber security, knowledge about the sources and underlying reasons of human failure and performance is still relatively scarce and "work in progress", but the amount of systematic research in this field is growing (Sütterlin et al. 2019).

Amongst the various ways in which human failure occurs and threatens cyber security, procedural compliance is one. A lack of procedural compliance when technology users do not adhere to existing security protocols can result from such protocols being too complicated, formulated in a, not understandable way, and highly technical. Other factors include the failure of organizational cultures, where enforced hierarchies and authoritarian leadership styles foster a culture of low tolerance to criticism and constructive feedback upwards along the vertical axis, making available competencies of lower ranking and technologically savvy younger experts unavailable for strategic decisions (Jøsok et al. 2016). Communicative challenges between individuals, teams, organisations and sectors particularly in interdisciplinary groups set up in an ad hoc manner and without established collaborative routines provide an environment that is particularly prone to misunderstandings, misinterpretations, and the loss of relevant information, resulting in discrepant mental models of a cyber threat situation. The reason why cyber security is particularly vulnerable to communicative challenges is the highly technical nature of the threat. A profound understanding and interpretation of a technical situation, its real-life consequences, options to act and their various anticipated consequences and associated risks in a cyber threat or cyber incident situation, with high stakes, that is potentially also characterized by time pressure and based on ambiguous information, creates an enormous cognitive individual and organizational workload prone to human errors. The creation of a shared situational awareness between technicians, decision-makers, and all other stakeholders requires a high degree of technical specialization in combination with efficient communication routines across disciplines, departments, organisations and societal sectors. The availability of understandable, yet accurate, technical information providing this shared situational awareness and decision-making ability gives technologically less informed decision-makers the ability to react and act with appropriate tactical and strategic decisions. The challenge is to provide an accurate yet simplified description of a given threat situation. The consequential options to act as well as the probabilities of anticipated consequences pose particular challenges in a context where not the shortage of information, but the overwhelming availability of it, adds to an enormous cognitive load for all stakeholders. In the communication of cyber threats, it is therefore not the situational status per se, but the outcome of its perception, interpretation, and communication by the cyber security technician that shapes the decision-makers' experienced reality. The decision-maker on the receiving end depends upon their understanding and thus upon the technician's simplification, selection, weighting, and interpretation. Knox et al. (2018) described the conditions that need to be put in place to facilitate information exchange on recognized cyber pictures between individuals and propose an orient-locate-bridge-model (OLB) describing how institutions can apply educational methods to enable both their cyber security personnel as well as their leaders in the effective communication of cyber security related situations (Knox et al. 2018).

Lacking procedural compliance and ineffective communication only represent examples of a number of ways in which human failures affect cyber security in and beyond the healthcare system. While there is a lack of research on human factor-related cyber security risks in the healthcare sector, the more extensive human-factor related knowledge from other domains can be transferred and applied in the healthcare sector as well. Human factor research can contribute to risk identification and the development of training approaches to facilitate cyber resilience. With 94% of malware distributed via email in 2018 (Verizon Communications Inc. 2019), phishing mails are the tool of choice to breach targeted networks. Empirical research has identified four categories of risk factors for susceptibility to phishing mails. These are situational factors, social engineering techniques, cultural factors, and individual differences (Canham et al. 2019). Employees who might not, under "normal" circumstances, be susceptible to a phishing attacks may be susceptible when distracted, or under significant cognitive load; in this way, situational factors play a major role in susceptibility. The social engineering techniques employed by threat actors can be very sophisticated and even appear to originate from known contacts. Some research suggests that cultural influences along the individualism–collectivism spectrum may significantly contribute to susceptibility (Butavicius et al. 2016). Finally, individual differences relating to personality and propensity to trust appear to account for some users being more susceptible than others. The most vulnerable users, sometimes referred to as "repeat clickers", represent a small minority of users who repeatedly fall prey to simulated phishing campaigns meant as training exercises. While these users usually only account for one to two percent of users within an organization, they can represent nearly 50% of the total simulated phishing failures (Canham et al. 2019).

Health care workers seem to fulfill some of these criteria, as they have shown a high propensity to click on phishing scams and have limited awareness of threats (Priestman et al. 2019). In a sector where women make up over 70% of workers in the health sector (WHO 2019) and usually rank higher on traits such as neuroticism, agreeableness (Parrish et al. 2009; Weisberg et al. 2011) and reward-based decision-making, known risk factors for maladaptive cyber behaviours cumulate. Getting targeted training to health care workers is essential in establishing and maintaining cyber resilience on an organisational level (Ghernaouti-Helie 2013). Other influencing factors determining susceptibility to fraudulent emails serving as an attack strategy to achieve entrance into sensitive IT systems are personality factors. IT users who are agreeable, emotionally less stable, and technically less knowledgeable, show higher risks of all unintended security violations in phishing attacks (Gratian et al. 2018; Halevi et al. 2013).

These vulnerabilities are usually addressed in cyber security education and training programmes. While these trainings are by far not sufficiently available and widespread in the healthcare sector, they are not even suitable to address all sources of errors. Even though general security training has shown to increase pro-security behaviors to some extent (Darwish et al. 2012), more targeted education is required to reach a significant improvement and lasting effects on behaviour.

In 2018, the United Kingdom's Information Commissioner (ICO) commissioned Kroll to conduct a study of all data breaches experienced by the UK government in 2017, 2124 incidents in total (Targett 2018). This study found that 88% of all data breaches involving UK government entities resulted from unintentional human error, without the direct involvement of a cyber threat actor. Examples of these errors included emailing unencrypted confidential patient information to the wrong recipient (21%), loss or theft of paperwork (20%) and data left in an insecure location (7%). Of the reported malicious cyber breaches, unauthorised access was the most common, accounting for 4% of the total, followed by malware (2.5%), phishing attacks (2.4%) and ransomware (1.5%). The healthcare sector accounted for the majority (57%) of these breaches, followed by general business (17%), education and childcare (16.7%), and local government (15.4%).

Knowledge workers in the information economy are often overworked, facing staffing shortages and constant deadlines. These workers commonly experience a tension between complying with security policy friction and accomplishing assigned tasks within deadlines (Posey and Canham 2018). Workers in the health care sector are no exception to this situation. One of the leading causes of (or contributing factors to) human error, is time pressure. Time pressure manifests in errors in two primary ways, deliberate policy violations and unintentional human errors (Reason 1990; Norman 2013).

The two categories of deliberate policy violations are routine violations and situational violations. Routine violations occur when policy non-compliance is so common that it is mostly ignored. A frequent example of this is the emergence of "shadow IT" systems. Shadow IT systems represent workarounds that users adopt in order to complete their work in an easier (but often less secure) manner. Examples include using unauthorized external cloud services or installing unauthorized wireless networks in secured spaces. Situational violations occur during exceptional circumstances. For example, in order to save a patient's life, emergency department medical staff might leave a proximity card on a monitoring cart in order to prevent the lock screen from activating. These violations are not routine, and usually occur when the cost of following proper security procedures is higher than abiding by them. Both routine and situational deliberate policy violations should give security staff pause, to reflect on the appropriateness of the policies causing these circumstances. If a policy is so cumbersome as to lead to it being deliberately ignored, this may be a good area to explore policy alternatives as a way to encourage secure behaviors. If existing IT infrastructure is so cumbersome to use that users are leveraging external resources, perhaps the existing infrastructure can be made more user friendly. Devices that utilize timeout lock screens might be modified to include an "emergency mode" so that they will not lock for the duration of the emergency.

Human errors, in contrast to deliberate violations, are unintentional and usually result from two distinctly different sources. The first type of human error is known as slips. Slips occur when an individual intends for one action to occur, but instead executes a different action. In this case, the person understands the correct action to take, but inadvertently takes the wrong action. An example of a slip might be pouring milk into coffee, but then placing the coffee cup in the refrigerator instead of replacing the milk (Norman 2013). In a cyber context, email address auto-complete is likely responsible for a great number of slip errors. Recall that the Kroll study found that emailing unencrypted confidential patient information to the wrong recipient was responsible for 21% of the 2017 data breaches (the largest number). In contrast to slips, mistakes are another type of human error that occur when an individual has the wrong goal (Norman 2013). Humans form mental models for how tools, artifacts, and environment operate and interact (Johnson-Laird 1983). These mental models encapsulate a simplified mental representation that allows the human mind to make predictions for actions taken with the things we interact with. Unfortunately, these mental representations are not always correct. When they are incorrect, this often leads to the wrong goal being formed, and an error then usually occurs. If a user believes that their employer has stronger security than what they have at home, they might forward a questionable email that they receive at their personal email account, to their work account, with the belief that the organization's security resources are better equipped to manage malicious software than their home system is. If this were the case, then this person would be taking the correct action, but of course this is false and based on an incorrect mental model.

This distinction between error types comes with consequence. In the cyber security industry, much emphasis is placed on user awareness training, with the implication being that if users were simply more aware of the inherent risks associated with their actions, they would be less vulnerable to attacks or committing errors leading to breaches (Carpenter 2019). While this may be true in some cases such as mistake type errors, it is untrue with regard to slip type errors. Slip errors result from highly learned behaviors that have become automatic and usually occur without conscious processing. In fact, slip errors tend to be more common in expert users, because they are so familiar with these actions. No amount of awareness training will fix slip errors; the best method to deal with these will likely be better design of interfaces and processes. Mistake errors result from incorrect goals, usually derived from incorrect mental models. Mental models can be updated through training and, therefore, these types of errors might be corrected through awareness training. Because these policy deviations can have radically different causes, they need to be addressed through different means (Canham et al. 2020). Security personnel in the health sector would be served well to track these errors in relation to data breaches and security violations and use these as a guide for developing corrective actions.

5. Cyber Resilience and Trust

In the previous sections, we laid out how digitalization transformed the way healthcare systems perform and the benefits these changes entail, as well as the vulnerabilities that come with large-scale health data administration. We also provided examples of how human behaviour can pose a large risk for breaches of sensitive data. Cyber resilience is more than the prevention of valuable data being stolen or the direct and collateral damages associated with a cyber attack. While data can in many cases be restored by backups, malware isolated and eliminated, and access to blocked data can be regained (ransomware)—there is a wider picture to it. Stolen data or access credentials and system vulnerabilities can spread in uncontrolled manners and be sold on illegal markets. The whereabouts of breached data remain usually unknown for long time periods, occasionally for years. The perpetrators can, in most cases, not be clearly identified, which adds an additional component of insecurity for the victims of cyber attacks, widely known as the "attribution problem". Cyber threats in the healthcare sector can have acute detrimental effects in times of national crisis (functionality of the healthcare system), hybrid warfare, or international conflicts above or below the threshold of war. In peacetime and in western democracies, however, breaches of data that were administered by private or public bodies (healthcare providers, insurances, etc.) can also undermine the public trust in these institutions. Public awareness and scepticism therefore influence policies around the digitalization of healthcare and consequently affects the development of institutions at the frontline of healthcare and achievement of sustainable development goals. The crucial role of people's trust in the protection of their privacy and thus in the integrity of the healthcare system as a whole has been recognized by state actors and lead to the development of relevant legal frameworks facilitating privacy, security and thus overall cyber resilience as a prerequisite of trust in a sustainable healthcare. According to the World Health Organization, trust in privacy legislation is key for the population's "confidence in their national eHealth programme" (WHO 2016, p. 77).

To reach a sufficient level of cyber resilience allowing for the further development, implementation and maintenance of digital solutions in healthcare, the vast majority of industrialized countries have established national legislation regulating the sharing, storing and use of personal data and/or personal health-related information as well as information exchange. While these legal frameworks have been put in place, the lack of cyber resilience on institutional levels makes healthcare institutions still an easy prey for malicious attacks or human failures due to mistakes and behavioural slips. One such cognitive barrier to prioritising cyber-resilience is the cost of investing in managing something that steals time from the primary and measurable role of the institution (Elgsaas and Heireng 2014). Investing time and resources into cyber secure systems and the people capable of ensuring persistent network resilience comes at considerable costs and is only rarely seen as a necessary and important investment in value generation (Coventry and Branley 2018). This is especially so when future funding for health institutions is very often performance related. Unfortunately, performance of cyber and information security is not the key criteria, meaning that in many countries, speed and efficiency of digital systems is prioritised ahead of security and resilience. For this reason, the introduction and implementation of digital systems in healthcare with the security by design principle at the forefront of development, and meet certain recognized classification frames, should be a step in adding cyber resilience from the start.

In Europe, national legal frameworks are directly affected by supernational legislation. The European Union Cyber Security Act of 2019 aims to increase EU resilience and response to cyber-attacks. The act established an EU framework for cybersecurity certification aimed at boosting the cybersecurity of digital products and services in Europe, including the various national healthcare sectors. In practice, the certification framework means improved cybersecurity across a wide spectrum of existing digital products and services, including the Internet of Things, as well as critical infrastructure such as, for example, hospitals. Even though these measures are not specifically aimed at cyber resilience in the healthcare sector, it contributes to the harmonization of cybersecurity standards, increasing the effectiveness in responding to cyberattacks as the system or device has a familiarity to it based on it meeting prescribed criteria. Manufacturers of healthcare technology are incentivised to invest in cyber security for their products, consequently giving them a potential competitive edge as customers see that certification is dependent upon a security-by-design approach to product development. When considering how the aforementioned potential of telemedicine and eHealth will rely heavily on the Internet of Things and the newest 5G data transmission standards, these new supernational legal frameworks can add a significant level of resilience; as it takes a stride in ensuring good cyber security as the foundation for trust in digital systems that should be able to guarantee information security.

Until legal frameworks such as the aforementioned legislation on harmonized certification have been implemented, security concerns remain a relevant factor influencing the users' trust in the healthcare system's ability to protect their sensitive data. The rise and investment in telemedicine to take advantage of the internet in support of ambulatory, more self-responsible monitoring and treatment in a "hospital-at-home" is heavily reliant upon security-by-design solutions minimizing or excluding the possibility of data breaches caused by unaware end-users. Such user-driven, self-monitoring of health care is ideal from a cost saving efficiency perspective as it is designed for people in certain risk-groups, people that might otherwise be re-hospitalized, and to assist people who are more likely to recover faster in their own home, whilst providing autonomy and enhanced quality of life. However, patients are required to place complete trust into their own network and the integrity of the data presented to them, or that they are required to share; as well as ensuring that the precise data that they need are available to them, their devices as well as to their healthcare professional, whenever it is needed and in real time. Lastly, the patient data must remain confidential in accordance with legal requirements. These demands are dependent upon a stable information and communication technology platform. Currently, many experts regard the Internet as simply not yet good enough for eHealth due to persisting reliability, availability and security issues. Each of which can undermine a patients' wellness due to the explicit risk related to telemedicine application and its dependency upon data management and data security.

Where even unintended individual human errors (slips) can cause massive data breaches, sensitive data of high value administered by poorly prepared and insufficiently aware healthcare professionals pose a huge incentive for malicious cyber attacks. The resulting limited trust in the system's integrity directly impairs the political incentives to facilitate digitalization further and increases the economic and organizational costs of further establishing robust cyber resilience. Precautions to ensure cyber resilience slow down or functionally impair the overall service performance by patients, due to delays, less user-friendly interfaces and identification requirements. As a result of diminished trust into both privacy and security, users may only reluctantly share highly personal data such as stigmatizing mental and sexual health conditions with a depersonalized and intransparent environment (Shenoy and Appel 2017). There is, to the best of our knowledge, currently no evidence-based knowledge on how far cyber security impairs the trust into the healthcare system and how reduced trust into cyber resilience impairs its performance or cost-effectiveness. Considering these likely side effects of hastily implemented digitalization without parallel implementation of a robust, transparent and technologically advanced cyber security strategy, we find the transparent, user-friendly and so far outstandingly successful cyber resiliency strategy provided by the Estonian example as a promising blueprint for future developments in other countries.

6. Conclusions

The digitalization of healthcare environments is changing the way healthcare systems operate and how they are organized, with significant changes for the patient's roles, responsibilities and opportunities. The benefits are manifold and relate to improved decision-making processes, availability of services, cost-effectiveness, and patient autonomy. The foundation of these changes is the constant and instant availability and exchange of patient- and service-related data that coordinate actors, communicate health-related patient data, provide the foundations for decision-making and facilitate the administrative processes at large. The comprehensive datasets occurring in this process provide the basis of the benefits of digitized healthcare, as well as the major challenge and vulnerability. A series of massive data breaches and enormous growth rates of cybercrime targeting valuable data processed in the healthcare systems, as well as a number of unintended breaches resulting from human failures, demonstrate an overproportional vulnerability of this societal sector, in which cyber security awareness is considered to lag behind other areas where security concerns appear more intuitively natural.

The health sectors' drive for digitalization to realize opportunities in, for example, eHealth and telemedicine will require far greater investment in cyber resilience if availability and security are to match the potential and ambitions of efficiency and effectiveness. Any progress towards ensuring the digitalization of the health sector needs to be measured against the current vulnerabilities to confidentiality, integrity and availability of before, between and after treatment data. Critical applications that are implemented to support achieving sustainable health goals may lack sufficient trust and reliability. Contact-tracing applications as they are introduced in the later phase of the COVID-19 pandemic have raised privacy and security concerns. While it is currently too early to thoroughly evaluate the outcomes of this particularly controversial symbol of accelerated digitalization in terms of public health benefits, the controversy around contact tracing apps provides an impressive example of the necessity of trust and its tight relationship with healthcare outcomes.

We argue that trust in healthcare systems affects its performance in a number of ways: (a) Patients hesitate to share sensitive data on their personal conditions if these are likely to be exchanged or administered in online databases and possible accessed by third parties. (b) The enormous additional investments in cyber security measures such as security-by-design or the engagement of qualified cyber security professionals as well as the related infrastructure increasing costs, planning and implementation time for digital innovations and thus slows down the transformation process. (c) Educational efforts by users including healthcare professionals, healthcare administrators, patients and third parties (for example, insurers) are necessary to develop cyber hygiene and reduce the risk of human failures, adding further to the time and financial costs of digitalized services.

Cyber security is fundamentally a human factor challenge and will require significant investment and research into achieving ways to develop a shared understanding across legal, institutional and national boundaries. The security aspect needs to be incorporated in the very early stages of designing and making it a central part of all digitalization processes.

While these measures to be taken may facilitate cyber resilience, increase the trust into a digitized healthcare with increased patient autonomy counteracting the simultaneous risk of privacy threats, parallel efforts have to be undertaken to ensure that all societal groups benefit from a more digitized healthcare. The empowerment and patient autonomy that comes with digitalization is of particular advantage for patients who know how to make use of their opportunities. Those with less technical affinity, but empowered, cannot take responsibility for their data security. This requires a balance of giving users full control and oversight over the use of their sensitive data, but without giving them the responsibility of making decisions that could unintentionally compromise their privacy.

In sum, the digitalization of healthcare is a potential major breakthrough in the development of sustainable healthcare worldwide. Sustainable healthcare system transformation, however, builds on the trust of its users and all measures taken to further improve the systems' effectiveness are subject to cost-benefit-analysis. Cyber resilience of healthcare systems plays an important role in building and ensuring ongoing trust as a central pillar of sustainability.

Author Contributions: All authors contributed substantially to all phases of the writing process, have approved the submitted version and agree to be personally accountable for the contribution as regards the accuracy and integrity of any part of the work.

Funding: This work was funded by the Research Council of Norway (Project nr. 302941).

Conflicts of Interest: The authors declare no conflict of interest.

References

Atasoy, Hilal, Brad N. Greenwood, and Jeffrey Scott McCullough. 2019. The digitization of patient care: A review of the effects of electronic health records on health care quality and utilization. *Annual Review of Public Health* 40: 487–500. [CrossRef] [PubMed]

Butavicius, Marcus, Kathryn Parsons, Malcolm Pattinson, and Agata McCormac. 2016. Breaching the human firewall: Social engineering in phishing and spear-phishing emails. *arXiv* arXiv:1606.00887.

Canham, Matthew, and Ben. D. Sawyer. 2019. Neurosecurity: Human Brain Electro-Optical Signals as MASINT. *American Intelligence Journal* 36: 40–47.

Canham, Matthew, S. Fiore, and B. Caulkins. 2018. Training Research to Improve Cyber Defense of Industrial Control Systems. In *Proceedings of the 62nd Annual Meeting of the Human Factors and Ergonomics Society*. Santa Monica: Human Factors and Ergonomics Society.

Canham, Matthew, M. Constantino, I. Hudson, S. M. Fiore, B. Caulkins, and L. Reinerman-Jones. 2019. The Enduring Mystery of Repeat Clickers. Paper presented at Fifteenth Symposium on Usable Privacy and Security (SOUPS 2019), Santa Clara, CA, USA, August 11–13; San Jose: USENIX.

Canham, Matthew, P. Bockelman, and C. Posey. 2020. To Train, Or Not to Train: Exploring the Boundaries of Security Education and Training Awareness. In *Lecture Notes in Computer Science*. New York: Springer.

Carpenter, Perry. 2019. *Transformational Security Awareness: What Neuroscientists, Storytellers, and Marketers Can Teach Us about Driving Secure Behaviors*. Hoboken: Wiley.

Chen, Qian, and Robert A. Bridges. 2017. Automated behavioral analysis of malware: A case study of wannacry ransomware. Paper presented at 2017 16th IEEE International Conference on Machine Learning and Applications (ICMLA), Cancun, Mexico, December 18–21; pp. 454–60.

Chien, Eric, Liam OMurchu, and Nicolas Falliere. 2012. W32. Duqu: The precursor to the next stuxnet. In *Presented as Part of the 5th {USENIX} Workshop on Large-Scale Exploits and Emergent Threats*. San Jose: USENIX Advanced Computing Systems Association.

Conrad, Eric, Seth Misenar, and Joshua Feldman. 2012. *CISSP Study Guide*. Rockland: Syngress.

Coventry, Lynne, and Dawn Branley. 2018. Cybersecurity in healthcare: A narrative review of trends, threats and ways forward. *Maturitas* 113: 48–52. [CrossRef] [PubMed]

Cusack, Brian, Kaushik Sundararajan, and Reza Khaleghparast. 2017. Neurosecurity for brainware devices. Paper presented at 15th Australian Information Security Management Conference, Perth, Australia, December 5–6.

Darwish, Ali, Ahmed El Zarka, and Fadi Aloul. 2012. Towards understanding phishing victims' profile. Paper presented at 2012 International Conference on Computer Systems and Industrial Informatics, Sharjah, UAE, December 18–20; pp. 1–5.

Department of Homeland Security (DHS). 2019. Available online: https://www.dhs.gov/topic/critical-infrastructure-security (accessed on 16 May 2020).

Elgsaas, Ingvill M., and Hege S. Heireng. 2014. Norges Sikkerhetstilstand–en Årsaksanalyse av Mangelfull Forebyggende Sikkerhet. Available online: https://www.ffi.no/no/Rapporter/ 14-00948.pdf (accessed on 15 May 2020).

Ghernaouti-Helie, Solange. 2013. *Cyber Power: Crime, Conflict and Security in Cyberspace*. Boca Raton: CRC Press.

Gratian, Margaret, Sruthi Bandi, Michel Cukier, Josiah Dykstra, and Amy Ginther. 2018. Correlating human traits and cyber security behavior intentions. *Computers & Security* 73: 345–58.

Halevi, T., J. Lewis, and N. Memon. 2013. A pilot study of cyber security and privacy related behavior and personality traits. Paper presented at ACM 22nd International Conference on World Wide Web, Rio de Janeiro, Brazil, May 13–17; pp. 737–44.

Health and Welfare Information Systems Centre (TEHIK). 2019. TEHIK—Data Security. Available online: https://www.tehik.ee/tervis/patsiendile/andmete-turvalisus/ (accessed on 30 November 2019).

Hughes, O. 2018. Norway Healthcare Cyber-Attack Could Be Biggest of Its Kind. *Digital Health*. Available online: https://www.digitalhealth.net/2018/01/norway-healthcare-cyber-attack-could-be-biggest/ (accessed on 21 February 2018).

Irwin, Luke. 2018. Breach at Norway's Largest Healthcare Authority Was a Disaster Waiting to Happen. *IT Governance Blog*. February 1. Available online: https://www.itgovernance.eu/ blog/en/breach-at-norways-largest-healthcare-authority-was-a-disaster-waiting-to-happen (accessed on 15 May 2020).

Jalali, Mohammad S., and Jessica P. Kaiser. 2018. Cybersecurity in hospitals: A systematic, organizational perspective. *Journal of Medical Internet Research* 20: e10059. [CrossRef] [PubMed]

Johansson, Grace. 2018. Half of Norway's Population Have Medical Data Leaked. *Cyber-Security News, Reviews and Opinion*. *SC Media UK*. January 19. Available online: https://www.scmagazineuk.com/half-norways-population-medical-data-leaked/ article/1473442 (accessed on 15 May 2020).

Johnson-Laird, Philip Nicholas. 1983. *Mental Models: Towards a Cognitive Science of Language, Inference, and Consciousness*. Cambridge, MA: Harvard University Press.

Jøsok, Øyvind, Benjamin J. Knox, Kirsi Helkala, Ricardo G. Lugo, Stefan Sütterlin, and Paul Ward. 2016. Exploring the hybrid space. In *Lecture Notes in Computer Science*. New York: Springer, vol. 9744, pp. 178–88.

Kim, Lee. 2017. Cybersecurity awareness: Protecting data and patients. *Nursing Management* 48: 16–19. [CrossRef] [PubMed]

Knox, Benjamin J., Øyvind Jøsok, Kirsi Helkala, Peter Khooshabeh, Terje Ødegaard, Ricardo G. Lugo, and Stefan Sütterlin. 2018. Socio-technical communication: The hybrid space and the OLB model for science-based cyber education. *Military Psychology* 30: 350–59. [CrossRef]

Kruse, Clemens Scott, Benjamin Frederick, Taylor Jacobson, and D. Kyle Monticone. 2017. Cybersecurity in healthcare: A systematic review of modern threats and trends. *Technology and Health Care* 25: 1–10. [CrossRef] [PubMed]

Lai, Taavi, Triin Habicht, Kristiina Kahur, Marge Reinap, and Raul Kiivet. 2013. Estonia: Health system review. *Health Systems in Transition* 15: 1–196. [PubMed]

Martin, Guy, Paul Martin, Chris Hankin, Ara Darzi, and James Kinross. 2017. Cybersecurity and healthcare: How safe are we? *BMJ* 358: j3179. [CrossRef] [PubMed]

McCoy, Thomas H., and Roy H. Perlis. 2018. Temporal trends and characteristics of reportable health data breaches, 2010–2017. *JAMA* 320: 1282–84. [CrossRef] [PubMed]

Norman, Donald A. 2013. *Design of Everyday Things: Revised and Expanded.* New York: Hachette.

Parrish, James L., Jr., Janet L. Bailey, and James F. Courtney. 2009. *A Personality Based Model for Determining Susceptibility to Phishing Attacks.* Little Rock: University of Arkansas, pp. 285–96.

Passeri, Paolo. 2017. May 2017 Cyber Attacks Statistics. *Hackmageddon.* July 13. Available online: https://www.hackmageddon.com/2017/07/13/may-2017-cyber-attacks-statistics/ (accessed on 15 May 2020).

Posey, C., and M. Canham. 2018. A Computational Social Science Approach to Examine the Duality between Productivity and Cybersecurity Policy Compliance within Organizations. Paper presented at 2018 International Conference on Social Computing, Behavioral-Cultural Modeling & Prediction and Behavior Representation in Modeling and Simulation (SBP-BRiMS 2018), Washington, DC, USA, July 10–13.

Priestman, W., T. Anstis, I. G. Sebire, S. Sridharan, and N. J. Sebire. 2019. Phishing in healthcare organisations: Threats, mitigation and approaches. *BMJ Health & Care Informatics* 26: e100031.

Priisalu, Jaan, and Rain Ottis. 2017. Personal control of privacy and data: Estonian experience. *Health and Technology* 7: 441–51. [CrossRef] [PubMed]

Reason, James. 1990. *Human Error.* Cambridge: Cambridge University Press.

Shenoy, Akhil, and J. M. Appel. 2017. Safeguarding confidentiality in electronic health records. *Cambridge Quarterly of Healthcare Ethics* 26: 337–41. [CrossRef]

Sütterlin, Stefan, Benjamin J. Knox, and Ricardo G. Lugo. 2019. TalTechi Küberkaitseteadlased: Ainult Tehnilistest Oskustest Ja Teadmistest Küberkaitses Ei Piisa. *Edasi.org.* October 30. Available online: https://edasi.org/48421/taltechi-kuberkaitseteadlased-ainult-tehnilistest-oskustest-ja-teadmistest-kuberkaitses-ei-piisa/ (accessed on 16 May 2020).

Targett, Ed. 2018. Revealed: Human Error, Not Hackers, to Blame for Vast Majority of Data Breaches. *Computer Business Review.* September 3. Available online: https://www.cbronline.com/news/kroll-foi-ico (accessed on 16 May 2020).

Verizon Communications Inc. 2019. *Verizon 2019 Data Breach Investigations Report.* New York: Verizon Communications Inc.

VG. 2017. 1200 Datamaskiner i Helse Sør-Øst Benytter Seg Av Utdatert Operativsystem. *Verdens Gang*. Available online: https://www.vg.no/nyheter/innenriks/i/BpbeQ/1200-datamaskiner-i-helse-soer-oest-benytter-seg-av-utdatert-operativsystem (accessed on 30 November 2019).

Weisberg, Yanna J., Colin G. DeYoung, and Jacob B. Hirsh. 2011. Gender differences in personality across the ten aspects of the Big Five. *Frontiers in Psychology* 2: 178. [CrossRef]

World Health Organization (WHO). 2016. From Innovation to Implementation—eHealth in the WHO European Region. Available online: http://www.euro.who.int/en/health-topics/Health-systems/e-health/publications/2016/from-innovation-to-implementation-ehealth-in-the-who-european-region-2016 (accessed on 16 May 2020).

World Health Organization (WHO). 2019. WHO Health Workforce—Data and Statistics. April 12. Available online: https://www.who.int/hrh/statistics/en/ (accessed on 12 April 2019).

Zhang, Xiang, and Badee uz Zaman. 2019. Adoption mechanism of telemedicine in underdeveloped country. *Health Informatics Journal* 26: 1088–1103. [CrossRef] [PubMed]

Sustainable Work Ability during Midlife and Old Age Functional Health and Mortality

Subas Neupane and Clas-Håkan Nygård

1. Introduction

Because of changes in demographics, the world is facing the challenge of aging societies, which can cause financial burden, as they generate less income for health services and pension systems, as well as a shortage of hands for production and taking care of the elderly. To overcome this problem, many national policies aim to further extend working careers by increasing the statutory retirement age. Extended working careers mean longer labor market participation and longer work exposure. However, if people should be in the labor market for a longer period, they ought to be in good physical and mental health and have access to more flexible working arrangements, healthy workplaces, and lifelong learning and retirement schemes. The process of living and working conditions that enable people to continue working for a longer period is here called sustainability (Eurofound 2015). Good health and wellbeing throughout life is one of the 17 sustainable goals of the United Nations (UN 2015), which ensure a healthy working life and promote the wellbeing of people at all ages. This means one of the key themes addressed by this SDG target is promoting health and safety in the workplace to ensure a healthy working environment and, thus, healthy lives during working careers and beyond by decent work, employment creation, social protection, and rights at work for productive employment and sustainable economic growth. Job quality and good working conditions contribute to wellbeing and a good quality of working life (Drobnič et al. 2010). However, workplace health and working conditions in most developing countries are generally poor in many work sectors, which means that the overall subjective wellbeing is poor in these countries (Robertson et al. 2016; Eurofound and ILO 2019).

Quality of life is one of the dimensions of sustainable work and is perceived to decrease by age; however, when controlled for potential factors, the effects of age may disappear. Sustainable work is an important topic worldwide. Several indicators, such as sustainable employment, employability, work engagement, and organizational commitment have been used in the literature to describe sustainable work concepts (Ilmarinen et al. 2005; Van Dam et al. 2017). However, making sustainable work requires a balance between the requirements of work, such as demands and control, and the needs of individuals, such as an individual's capacity, as both change over time. Moreover, whether a healthy and longer work-life is possible for every person remains unclear (Ilmarinen 2013). Good health among individuals is a prerequisite for this continuation, but knowledge about the possible health consequences of an extended working life is limited. Earlier research shows that retirement rather than an extended working career tends to be a relief for those who suffered from suboptimal self-rated health, sleep disturbances, fatigue, depression or headache diseases (Westerlund et al. 2009; Stenholm et al. 2016), or musculoskeletal pain (Neupane et al. 2018), whereas retirement seems to have an immediate positive effect on the risk of the incident of chronic conditions. Other studies have shown that exposure to poor working conditions at earlier stages of the working life may increase the possibility of poor health and disability in later life. The number of risk factors of physical functioning in later life start to accumulate from early to midlife, and include behavioral, environmental, lifestyle-related, and sociofactorial factors (Guralnik et al. 1993; Brown and Flood 2013). Yet, only a little is known about the link between work-related exposures from midlife and functional health in later life. There exists some evidence on the negative association between occupational physical activities (Hinrichs et al. 2014) and work stress (Verbrugge and Jette 1994), as well as work ability (Kulmala et al. 2014) in midlife and mobility limitations in later life. Physically demanding work with vigorous occupational physical activity in midlife increases the risk of mobility limitations in later life (Hinrichs et al. 2014), which could eventually progress to disability. Similarly, lower work-related stress (Kulmala et al. 2014) and better work ability (von Bonsdorff et al. 2016) in midlife have been reported to be protective to mobility limitations in later life, while poor work ability was reported to be associated with mortality (von Bonsdorff et al. 2011).

The occupation in midlife also plays a vital role in the onset of later life disability, with unskilled blue-collar workers being the high-risk group compared to white-collar workers (Prakash et al. 2016). Work-related high physical exposure and job strain in midlife were strongly associated with the severity of disability in later life (Prakash et al. 2017).

Sustainable work ability is a multifaceted concept that involves the matching of the needs and abilities of the individual with the quality of jobs on offer. Although sustainable work ability is often mentioned in reference to an aging workforce, it should be noted that sustainability relates to workers of all ages (Van Vuuren and Van Dam 2013). In order to stay in the workforce until retirement, it is important that employees work in a healthy workplace, whatever their age.

Work ability, defined as people's ability to cope with their work demands, is a broad concept and an important human capital of workers throughout their working career (Ilmarinen 2019). It requires continuous processes at workplaces, aiming to improve the fitting of human resources and work environments together. Proper working conditions enable a good fit between work and the characteristics of the individual throughout their working life. To achieve this dual goal, employers are required to develop new solutions for working conditions and career paths that help workers to maintain their work ability over an extended working life. Work ability tends to decline with age (Ilmarinen Juhani and Matti 1997), although the mean work ability of the working population between the ages of 20 and 65 years remains at a good or excellent level (Gould et al. 2008). In this chapter, work ability maintained at good or excellent during the work career was considered sustainable work ability. We studied the impact of sustainable work ability by examining trajectories over 16 years. Wellbeing was measured in terms of mobility limitations in old age after 12 years using longitudinal data on employees in a large amount of blue- and white-collar municipal occupations. We also studied the difference in survival and any cause of death among people in work ability trajectory groups. Our hypothesis was that people with sustainable work ability during midlife would have less mobility limitations and better survival than those who did not have sustainable work ability.

2. Materials and Methods

The Finnish Longitudinal study on Aging Municipal Employees (FLAME) was conducted among municipal workers from 1981 to 2009 (Ilmarinen et al. 1991; Tuomi et al. 1997; von Bonsdorff et al. 2011). At baseline, in 1981, a postal questionnaire was sent to 7344 municipal workers all around Finland. In total, 6257 (85.2%) 44–58-year-olds having worked as municipal workers for at least 5 years responded. Follow-up data were collected with postal questionnaires in 1985 (n = 5556), 1992 (n = 4534), 1997 (n = 3815), and 2009 (n = 3093) (von Bonsdorff et al. 2011). In this study, we analyzed work ability data from 16 years of follow-up from 1981 to 1997, which covers the work career of people from midlife until retirement. For inclusion in the trajectory analysis, the respondent must have replied at baseline, while in the regression analysis, we analyzed only those participants who had information on the outcome variable (mobility limitations) from the last round (2009) of follow-up (n = 2918). The respondents' exact retirement dates and mortality dates, from any cause, were obtained from the national pension registry and linked with the survey data. Figure 1 presents the follow-up process in detail.

The Ethics Committee of the Finnish Institute of Occupational Health, Helsinki, Finland approved the study.

2.1. Measurement of Variables

2.1.1. Work Ability

In this analysis, we used two types of work ability measures from the work career. Work ability index (WAI) was the composite measure of seven items (Table 1) based on subjective survey instruments (Ilmarinen et al. 1991). WAI indicates the wellbeing, health status, and quality of life as it measures how good a worker is at present and in the near future, and ability to work with respect to work demands (Ilmarinen et al. 2005). WAI was measured only at baseline

Table 1. Items of work ability index.

Items	Range
1. Current work ability compared with the lifetime best	0–10
2. Work ability in relation to the demands of the job	2–10
3. Number of current diseases diagnosed by a physician	1–7
4. Estimated work impairment due to diseases	1–6
5. Sick leave during the past year (12 months)	1–5
6. Own prognosis of work ability 2 years from now	1–7
7. Mental resources	1–4

The WAI ranges from 7 to 49, and the higher the score, the better the work ability. Based on the scores, WAI has been classified into four standard categories (poor 7–27, moderate 28–36, good 37–43, and excellent 44–49) (Gould et al. 2008). In this analysis, good and excellent work ability was combined.

The work ability score (WAS) was measured using a single item; the first item of the WAI is "current work ability compared with the lifetime best" with a score from 0 to 10, where 0 is incapable to work and 10 is the work ability at its best (von Bonsdorff et al. 2011; Ebener and Hasselhorn 2019). WAS is strongly associated with WAI and can be used as a simpler indicator for assessing work wellbeing (Ahlstrom et al. 2010). WAS was measured identically in all four surveys. The developmental patterns of work ability from four time points were used as the main independent variable in this analysis.

2.1.2. Mobility Limitation

Mobility Limitations (MLs) as an outcome of interest in this study were measured using self-reported questionnaires distributed among the participants in the last round of the follow-up in 2009. The International Classification of Functioning, Disability, and Health (ICF) was used to define the mobility limitations (World Health Organization 2001). Nine items related to activities and the mobility of the participants were used to create a final score. Table 2 shows the items included in defining mobility limitations. All the items, except walking 2 km, were assessed on a four-point scale of difficulty (manage without difficulties, manage with little difficulties, manage with lots of difficulties, and cannot manage). Walking 2 km was assessed on five levels (no difficulty—cannot manage with the help of others as well). In this analysis, all nine items were first dichotomized (no difficulty vs. at least some difficulty), then combined to make a summary score of 0 to 9 (score '0' represented no limitations in carrying out any of the 9 tasks and those who had at least some limitations in carrying out one or more of the 9 tasks scored '1–9' depending on the number of tasks entailing limitations) (Prakash et al. 2019; Hinrichs et al. 2014).

Table 2. International classification of functioning (ICF), disability, and health categories used to create mobility limitation (ML) (0–9).

Mobility	List of Categories	Questions	Range
Changing and maintaining body position	Changing basic body position	1. Squatting and standing up again?	1–4
		2. Bending down deep?	1–4
	Maintaining a body position	3. Maintaining body position/sitting still for 2 h?	1–4
Carrying moving and handling objects	Lifting and carrying objects	4. Lifting and carrying more than 10 kg?	1–4
	Fine hand use	5. Precise movements of hands?	1–4
	Hand and arm use	6. Lifting hands over the head?	1–4
Walking and moving	Walking	7. Walking 2 km?	1–5
	Moving around	8. Running 100 m?	1–4
		9. Climbing three floors/stairs?	1–4

2.1.3. Mortality

The study participants were followed for mortality between 1 January 1981 and 31 July 2009. Data on date of death, from any cause, were obtained from the Finnish National Population Register.

Demographic Information

Age (44–58 at baseline) was used as a continuous and categorical (44–49 vs. 50–58 years) variable in the analyses. Occupational class (white-collar, blue-collar) was created based on a detailed analysis of job profiles among 88 occupational titles, clustered into 13 job profiles and later, into two major groups (Kulmala et al. 2014; Prakash et al. 2016). Information on gender (female/male) was obtained from the questionnaire survey.

Lifestyle Characteristics

Leisure-time physical activity (LTPA) in the past year was collected in five categories (1—brisk exercise at least twice a week; 2—brisk exercise at least once a week; 3—some exercise at least once a week; 4—some exercise less than once a week or no exercise) and classified as high (1–2) or low (3–4) (Neupane et al. 2018). BMI (kg/m^2) was calculated using self-reported height and weight and dichotomized as <25.0 (normal) or ≥25.0 (overweight/obese). Those reporting current smoking >1 cigarette per day or past smoking were classified as smokers.

Morbidity

Information on morbidity was obtained by the question "Please indicate in the list below which diseases or impairments you have at present. In addition, check whether a physician has diagnosed or treated this condition". The list covered 47 items. We used the following categories (yes/no) of physician-diagnosed diseases: musculoskeletal, cardiovascular, respiratory, and metabolic diseases. The information on the number of diseases was summed up and categorized into three—0, 1, 2, or more morbidities.

Physical Workload

Physical workload at baseline was assessed with eight questions about current exposure to the following: vibration, repeated movements, standing still, bent or twisted postures, other poor postures, continuous walking or movement, carrying objects, and sudden strenuous efforts. Response options ranged from 0 (not at all) to 4 (quite often). The composite score (Cronbach's $\alpha = 0.82$) ranging from 0 to 32 was dichotomized into high and low at the median value of 12, with median value included in high physical workload (Neupane et al. 2018).

2.2. Statistical Analyses

We first analyzed the mean distribution of the work ability index by age (as a continuous variable) using the box plot method. The cut-off line was plotted to separate out 'good and excellent', 'moderate', and 'poor' work ability.

Growth mixture modeling (GMM) was applied to identify trajectories of the work ability score (WAS) from four time points. The basic assumption of GMM is that all individuals follow the growth pattern of a random variation. GMM also accounts for within-class variations in the estimation of class memberships (Ram and Grimm 2009). All the study participants who responded at baseline to the survey were included in the trajectory analysis. The number of trajectories and their shape were determined first. The quadratic function best represented the patterns of change in the WAS. The final model was chosen based on a range of fit criteria (see appendix), including the Akaike Information Criterion (AIC), the Bayesian Information Criterion (BIC), sample size adjusted BIC, entropy, and posterior probabilities and meaning and their interpretability (Nylund et al. 2007). Based on these, a three-trajectory model was selected. The trajectory groups were illustrated by plotting the means of the WAS against survey year.

Baseline characteristics of the study participants were presented as frequencies and percentages by work ability trajectory. The difference between trajectories was tested by the Chi-square test.

Sixteen-year trajectories of the WAS were then used to study the associations with mobility limitations of the people after 12 years. A generalized linear model (GLM) with a Poisson function was used to calculate the incidence rate ratios (IRR) and their 95% confidence intervals (CIs) for the associations. Three models were fitted, the first model was the bivariate association of the trajectories of the WAS with mobility limitations. Bivariate association of the sociodemographic, behavior, and work-related variables from the baseline with the mobility limitations was also calculated in Model I. In Model II, the association of work ability trajectories with mobility limitation was adjusted for sociodemographic variables (age, gender, and occupational class). Model III was further adjusted for all the sociodemographic, lifestyle-related, morbidity, and work-related variables from Model I. The estimates with their 95% CIs are also presented for all the covariates used in the Model III.

Nelson–Aalen cumulative hazard estimates were plotted for mortality from any cause by trajectory membership. The follow-up time started from the baseline, 1 January 1981, and ended with censoring resulting from death or end of follow-up in 2009.

Trajectory analysis was conducted in Mplus v7.2 and the other analyses in R-studio and Stata v15.

3. Results

Figure 1 shows the mean distribution of work ability index at baseline by age of respondents. It shows that, in general, the work ability decreases with increasing age. The mean work ability was maintained as either good or excellent until the age of 51 years; after that, the mean level of work ability index decreased to moderate.

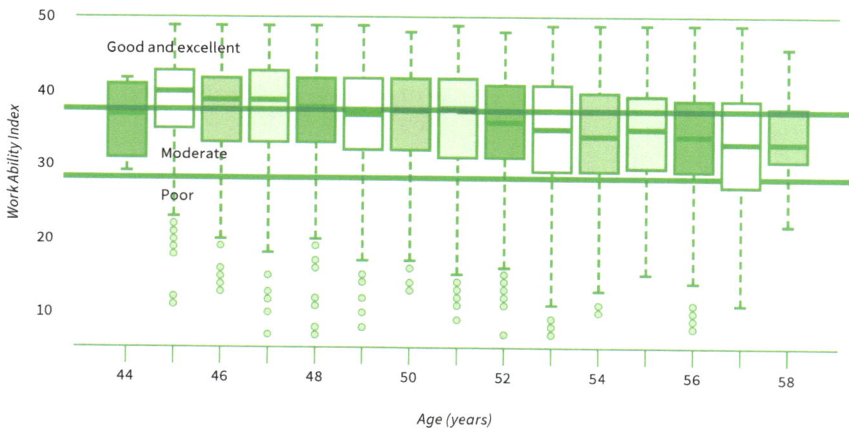

Figure 1. Distribution of the work ability index (7–49) by age (44–58 years) among study participants at baseline.

Participants were followed for 16 years after the baseline survey until 1997 to see the changes in WAS. We used trajectory analysis to see the change in work ability score using self-reported WAS data from four time points. Figure 2 shows the trajectories of mean WAS at different follow-up times. We found three distinct trajectories.

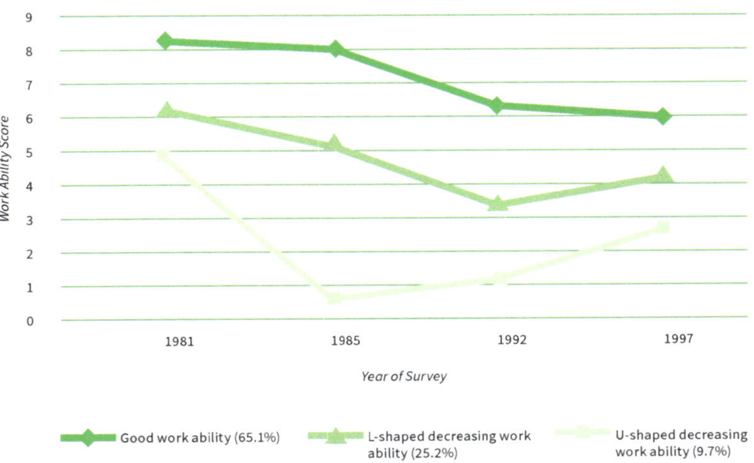

Figure 2. Trajectories of the mean of the work ability score from 1981 to 1997 among study participants.

Of the total subjects studied, the majority (65%) had good work ability during the 16 years of follow-up. The mean work ability score was about 8 at the baseline, which slightly decreased to, on average, 6 during the follow-up. This group of people who maintained good work ability during their work career was considered here as the sustainable work ability group. About 25% of the study participants had L-shaped decreasing work ability. They started with a mean WAS of 6 at baseline, which then later decreased in L shape during the follow-up. The third group consisted of about 10% of the participants, who had a U-shaped decreasing work ability. They started with a mean WAS of 5 at baseline, which decreased as U-shaped during the follow-up. In the good and L-shaped decreasing trajectories group, the decrease in the work ability score was steep until 1992; after that, the work ability mostly remained constant or improved. Meanwhile, among those in the U-shaped trajectory group, a slight improvement in work ability was observed already from 1992.

Table 3 shows the baseline characteristics of the study population by work ability trajectories. There were significantly more younger workers who had a good work ability trajectory compared to their older counterparts, while more older workers had L- or U-shaped decreasing work ability. Significantly more women, white-collar workers, high LTPA, low (<25) BMI, no or one comorbidity, and workers with low physical workload had good work ability. More men and more blue-collar workers had L- and U-shaped decreasing work ability. Meanwhile, ther was no significant difference in the distribution of work ability trajectories by smoking status.

Table 3. Baseline characteristics of the study sample by work ability trajectory groups.

Characteristics	Total N = 2918	Trajectory Membership			p-Value
		Good (n = 2083)	L-Shaped Decreasing (n = 661)	U-Shaped Decreasing (n = 174)	
Age (Years)					<0.001
44–49	1529	1149 (75.1)	319 (20.9)	61 (4.0)	
50–58	1389	934 (67.2)	342 (24.6)	113 (8.1)	
Gender					<0.001
Women	1858	1348 (72.5)	410 (22.1)	100 (5.4)	
Men	1060	735 (69.3)	251 (23.7)	74 (7.0)	
Occupational class					<0.001
White-collar	1688	1318 (78.1)	307 (18.2)	63 (3.7)	
Blue-collar	1230	765 (62.2)	354 (28.8)	111 (9.0)	
Smoking					0.447
No	1873	1345 (71.8)	424 (22.6)	104 (5.6)	
Yes	1045	738 (70.6)	237 (22.7)	70 (6.7)	
LTPA [†]					<0.001
High	1505	1130 (75.1)	294 (19.5)	81 (5.4)	
Low	1354	915 (67.6)	350 (25.9)	89 (6.5)	
BMI [‡]					<0.001
<25.0	1442	1091 (75.7)	283 (19.6)	68 (4.7)	
≥25.0	1449	976 (67.4)	368 (25.4)	105 (7.2)	
Comorbidity					<0.001
0	1337	1145 (85.6)	168 (12.6)	24 (1.8)	
1	810	557 (68.8)	202 (24.9)	51 (6.3)	
2 or more	771	381 (49.4)	291 (37.7)	99 (12.8)	
Physical workload					<0.001
Low	1464	1195 (81.6)	229 (15.6)	40 (2.7)	
High	1454	888 (61.1)	432 (29.7)	134 (9.2)	

[†] Leisure-time physical activity; [‡] Body mass index.

The distribution of mean value of mobility limitations with their 95% CIs by work ability trajectories is presented in Table 4. It shows that the mean number of mobility limitations was lowest (mean 3.38, 95% CI 3.27–3.49) among those in the good work ability trajectory group, while it was highest (5.51, 95% CI 5.13–5.90) among those in the U-shaped decreasing trajectory group. Similarly, in the same table (Table 4), the mean work ability score is presented by survey year. This shows that the mean score decreased by survey year because of the increase in age of the workers in each survey round.

Table 4. Distribution of the mean mobility limitations (0–9) according to work ability trajectory group and mean work ability score (0–10) by year of survey.

Work Ability	Mobility Limitation Mean, 95% CI
Good	3.38 (3.27–3.49)
L-shaped decreasing	4.71 (4.51–4.92)
U-shaped decreasing	5.51 (5.13–5.90)
Year of Survey	**Work Ability, Mean ± SD**
1981	7.65 ± 1.70
1985	6.93 ± 2.14
1992	5.62 ± 2.79
1997	5.65 ± 2.42

The associations of work ability trajectories with the mobility limitation are presented in Table 5. It shows strong and statistically significant associations of L-shaped and U-shaped decreasing work ability trajectories with increased mobility limitations compared to those belonging to the good work ability trajectory group, so-called people with sustainable work ability. The crude model (Model I) shows that belonging to the L-shaped and U-shaped decreasing work ability trajectory groups was associated with increased risk of mobility limitations in the 12 years follow-up compared to those in the good work ability trajectories group. The associations remained statistically significant even after adjusting for age, gender, and occupational class in Model II. After further adjustment with smoking status, LTPA, BMI, morbidity, and physical workload in Model III, the association remained strong and statistically significant (IRR for L-shaped decreasing work ability 1.24, 95% CI 1.18–1.30 and U-shaped decreasing work ability 1.37, 95% CI 1.28–1.47). The magnitude of the association was higher for those in the U-shaped decreasing trajectory group in each model.

Table 5. Associations of 16-year work ability trajectories with mobility limitations in 28 years follow-up. Incidence rate ratios (IRR) and their 95% confidence intervals (CIs) from a Poisson regression model.

Characteristics	N [†] = 2533	IRR, 95% CI for Mobility Limitation		
		Model I	Model II	Model III
Work ability				
Good	1751	1	1	1
L-shaped decreasing	615	1.40 (1.34–1.46)	1.37 (1.31–1.42)	1.24 (1.18–1.30)
U-shaped decreasing	167	1.65 (1.54–1.76)	1.57 (1.46–1.68)	1.37 (1.28–1.47)
Age (Years)				
44–49	1278	1	1	1
50–58	1260	1.28 (1.23–1.32)	1.23 (1.18–1.27)	1.19 (1.14–1.23)
Gender				
Women	1707	1	1	1
Men	826	0.80 (0.76–0.83)	0.76 (0.73–0.79)	0.73 (0.69–0.76)
Occupational class				
White-collar	1460	1	1	1
Blue-collar	1078	1.13 (1.09–1.18)	1.13 (1.09–1.17)	1.03 (0.99–1.08)
Smoking				
No	1639	1		1
Yes	899	1.00 (0.97–1.04)		1.11 (1.06–1.16)
LTPA [†]				
High	1252	1		1
Low	1233	1.25 (1.20–1.29)		1.16 (1.11–1.20)
BMI				
<25.0	1202	1		1
≥25.0	1313	1.32 (1.27–1.37)		1.25 (1.20–1.30)
Comorbidity				
0	1092	1		1
1	731	1.27 (1.21–1.33)		1.16 (1.11–1.22)
2 or more	715	1.53 (1.46–1.60)		1.28 (1.22–1.34)
Physical workload				
Low	1230	1		1
High	1308	1.27 (1.22–1.32)		1.08 (1.04–1.13)

[†] Participants with at least some mobility limitations. Model I: Bivariate model; Model II: Adjusted for age, gender, and occupational class; Model III: Adjusted for all variables from Model II + smoking, LTPA, BMI, comorbidity, and physical workload.

The increasing rate of mobility limitations was found to be associated with older age, those who never smoked, low LTPA, higher BMI, those with one or more comorbidities, and those who reported high physical workload. These associations remained statistically significant in the final model (Model III). Meanwhile, compared to women, men were associated with lower incidence rates of mobility limitations and the association of blue-collar workers with mobility limitations was not clear in the final model.

Figure 3 presents the Nelson–Aalen cumulative hazard estimates by work ability trajectories group for all-cause mortality. The estimates in the different trajectory group already started to diverge after 5 years from the baseline survey. The hazard was highest among those in the U-shaped decreasing work ability trajectory group and lowest among those in the good work ability trajectory group. The estimates in the good and L-shaped decreasing work ability trajectory group started to diverge after 10 years of follow-up.

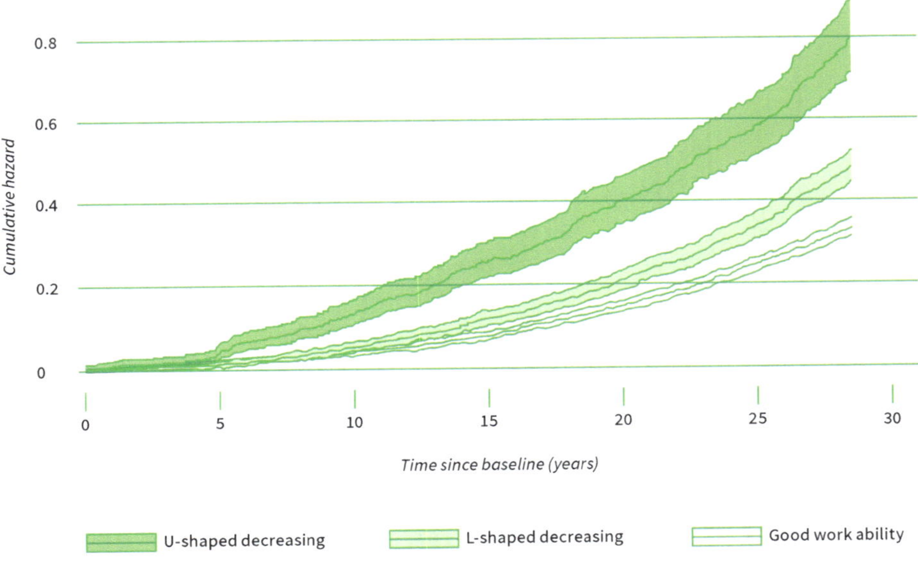

Figure 3. Nelson–Aalen cumulative hazard estimates of all-cause mortality by work ability trajectory group between 1981 and 2009.

4. Discussion

We used good work ability as an indicator of sustainable work ability in a long-term prospective follow-up study of municipal employees to study the health impact in later life. We found that about two-thirds of our study population had good or sustainable work ability during their late work career. Sustainable work ability was a very strong predictor of good health outcome in terms of mobility limitations in old age. People belonging to the trajectories of decreasing work ability had increased risk of mobility limitations in old age. They also had a high cumulative hazard of death from any cause (i.e., worst survival) compared to those with a good work ability trajectory. Overall, our findings supported the importance of a sustainable work ability during the latter part of the work career.

Work ability has been used as a continuity construct since its establishment in the Finnish Institute of Occupational Health (Tuomi et al. 1998; Ilmarinen 2009, 2019). It considers the surrounding organizational and societal context, providing a wide perspective on the relationship between the individual and the work and social environment (Ilmarinen 2009). Work ability has been supported by the goal to prolong work careers and to prevent work disability (Gould et al. 2008). Work ability is also associated with health-related quality of life (Sörensen et al. 2008), which indicates wellbeing at present and its sustainability. It is, therefore, important to maintain work ability during the work career for better quality of life at midlife and beyond.

In our study, work ability was fairly stable for about two-thirds of the people during 16 years of follow-up in their midlife and the rest had decreasing work ability until the first 11 years of follow-up andm, then, slowly improved after that. This indicates that the majority of the employees had sustainable work ability and wellbeing at midlife. The work ability index at baseline also shows that the mean work ability was maintained either at good or excellent until the age of 51 years; after that, the mean level of work ability index decreased to moderate. There were 20–25% of 44–49 years old who had moderate or poor work ability. They risk losing their work ability if no preventive actions are taken. The stability of work ability indicates that the employees have enough resources to cope with their job demands, although there is a wide variation between white- and blue-collar occupations but less variation by gender. The majority of our study participants (75%) retired during the second round of the follow-up in 1992, and by the end of 1997, almost all (99%) were retired. The decreasing work ability trajectory group improved their mean work ability score after 1992, which means that retirement improved their perceived work ability. This means the quality of life and wellbeing of people improves when there is no pressure of work demands.

These results of our study are consistent with the French GAZEL study, showing that perceived health in older workers, exposed to poor working conditions, is relaxed after retirement (Westerlund et al. 2009). Another study also showed the health benefits of retirement but only for upper occupational class employees (Mein et al. 2003). Employees will be relieved from their physical and mental demands of the work, which is beneficial to health and quality of life.

Consistent with our results, similar findings have been reported in earlier studies. A study from the USA reported three trajectories of work ability with 74% having good work ability, 17% declining, and only few, 9%, having poor work ability (Boissonneault and de Beer 2018). An earlier study, from our FLAME study but with a longer follow-up, until old age, reported five trajectories of work ability, with a substantial proportion of individuals maintaining their work ability at a moderate level (von Bonsdorff et al. 2011). Our earlier study among younger working aged (mean age 42 years) people in the manufacturing industry showed that 90% had a good work ability trajectory during six years of follow-up (Oakman et al. 2019).

Trajectories of decreased work ability in either an L- or U-shape were associated with increased risk of mobility limitations in old age, even after controlling for the effect of demographic, lifestyle factors, morbidity and physical workload in the baseline. Mobility limitations in this study were used as an indicator of functional health in old age. Mobility limitations are associated with poor quality of life among older adults and predict all-cause mortality (Bergland et al. 2017). Consistent with our findings, an earlier study reported that better work ability protects from old age mobility limitations among those who retire due to non-disability and disability (von Bonsdorff et al. 2016). Other studies found that vigorous occupational physical activity (Hinrichs et al. 2014), higher work stress (Kulmala et al. 2014), and shift work (Prakash et al. 2019) were associated with mobility limitations in old age. We also found that among studied covariates, older age, being a woman, smoker, low leisure-time physical activity, high BMI, having multiple morbidities as well as having high physical workload at baseline were statistically significantly associated with an increased risk of mobility limitations in old age. Moreover, higher cumulative risk of mortality from any cause was found among those in the decreasing trajectories group compared to those in the sustainable work ability group.

Sustainable work ability is beneficial to everyone, industries and society as a whole. Individuals can benefit from better work outcomes, smoother transitions between life stages, and longer working lives from the sustainable work ability, whereas, for industries, sustainable work ability may lead to an improved efficiency or productivity. This implies that society will benefit from healthier populations, higher employment rates, more inclusive labor markets, and lower pressure on public budgets. Due to the ageing of the population and the shrinking workforce, new sources of growth and economic progress are needed where the sustainable work ability concept would be relevant. Sustainable work ability promotes a holistic approach that considers workers' health, personal characteristics, family, and social responsibilities. Those with a diminishing work ability can be promoted for example, decreasing physical workload, adjusting work–rest schedules, and introducing age-management practices, flexible working time schedules, and teamwork (Ilmarinen and Rantanen 1999).

Sustainable work ability requires sustainable employment and wellbeing at work, which then enhances the quality of work life and satisfaction. The aging workforce also emphasizes the importance of sustainable employment; therefore, workers are willing to work after their normal retirement age (Armstrong-Stassen and Schlosser 2008; Gobeski and Beehr 2009), which can have the potential to achieve sustainable economic growth. "Promote sustained, inclusive and sustainable economic growth, full and productive employment and decent work for all" is one part of the agenda of the SDGs. Work organizations may have an important role, which can draw attention to an age-supportive atmosphere for workers of all ages (Oakman et al. 2016), although age-related changes are largely specific to individuals, with wide interpersonal variability (Sluiter 2006). A major strength of our study was the use of long prospective follow-up data from a representative population of Finnish municipal occupations. Among other strengths, there was a high response rate at baseline and follow-ups, use of official registers data on retirement and mortality, and elimination of recall bias. About 50% of the baseline respondents also participated in the last survey wave in 2009 after almost three decades. One can consider some of the limitations of this study while inferring the findings. The work ability scores as well as mobility limitations were both self-reported, which is possibly subject to information and recall bias and could lead to over-reporting. Nevertheless, these measures are validated tools and have been used extensively in earlier research (von Bonsdorff et al. 2011; Prakash et al. 2019; Hinrichs et al. 2014; Ilmarinen 2019). Another strength was that we used ICF classification of physical functioning that has been validated and frequently used in earlier studies (Hinrichs et al. 2014).

5. Conclusions

In all, we found that the majority of people maintain fairly good work ability during their work career in midlife, while a few have declining work ability. Sustainable work ability protects from mobility limitations in old age, while decreasing trajectories of work ability are associated with increased risk of mobility limitations as well as premature mortality. It is, therefore, important to detect workers with diminishing work ability in order to promote their work ability and prevent from future disability.

Author Contributions: C.-H.N. was involved in data collection of initial surveys. S.N. analyzed the data, interpreted results, and drafted the manuscript. S.N. and C.-H.N. critically read and revised the manuscript and accept the final version as submitted.

Funding: S.N. was partly supported by the Competitive State Research Financing of the Expert Responsibility area of Tampere University Hospital.

Acknowledgments: We would like to thank the Finnish Institute of Occupational Health for the opportunity to use their data, and especially Jorma Seitsamo for providing the data for the current study. We would also like to thank Juhani Ilmarinen for his support in planning of the manuscript.

Conflicts of Interest: The authors declare no conflict of interest.

References

Ahlstrom, Linda, Anna Grimby-Ekman, Mats Hagberg, and Lotta Dellve. 2010. The work ability index and single-item question: Associations with sick leave, symptoms, and health—A prospective study of women on long-term sick leave. *Scandinavian Journal of Work, Environment & Health* 36: 404–12.

Armstrong-Stassen, Marjorie A., and Francine Schlosser. 2008. Benefits of a supportive development climate for older workers. *Journal of Managerial Psychology* 23: 419–37. [CrossRef]

Bergland, Astrid, Lone Jørgensen, Nina Emaus, and Bjørn H. Strand. 2017. Mobility as a predictor of all-cause mortality in older men and women: 11.8 years follow-up in the Tromsø study. *BMC Health Services Research* 17: 22. [CrossRef] [PubMed]

Boissonneault, Michael, and Joop de Beer. 2018. Work Ability Trajectories and Retirement Pathways: A Longitudinal Analysis of Older American Workers. *Occupational and Environmental Medicine* 60: e343. [CrossRef] [PubMed]

Brown, Cynthia J., and Kellie L. Flood. 2013. Mobility limitation in the older patient: A clinical review. *JAMA* 310: 1168–77. [CrossRef] [PubMed]

Drobnič, Sonja, Barbara Beham, and Patrik Präg. 2010. Good job, good life? Working conditions and quality of life in Europe. *Social Indicators Research* 99: 205–25. [CrossRef]

Ebener, Melanie, and Hans Martin Hasselhorn. 2019. Validation of Short Measures of Work Ability for Research and Employee Surveys. *International Journal of Environmental Research and Public Health* 16: 3386. [CrossRef]

Eurofound and International Labour Organization. 2019. *Working Conditions in a Global Perspective*. Luxembourg: Publications Office of the European Union, Geneva: International Labour Organization.

Eurofound. 2015. *Sustainable Work over the Life Course: Concept Paper*. Luxembourg: Publications Office of the European Union, Available online: https://www.eurofound.eu ropa.eu/sites/default/files/ef_publication/field_ef_document/ef1519en.pdf (accessed on 8 December 2019).

Gobeski, Kirsten T., and Terry A. Beehr. 2009. How retirees work: Predictors of different types of bridge employment. *Journal of Organizational Behavior* 30: 401–25. [CrossRef]

Gould, Raija, Juhani Ilmarinen, Jorma Järvisalo, and Seppo Koskinen. 2008. *Dimensions of Work Ability: Results of the Health 2000 Survey*. Helsinki: Finnish Centre of Pensions (ETK), Helsinki: The Social Insurance Institution (KELA), Helsinki: National Public Health Institute (KTL), Helsinki: Finnish Institute of Occupational Health (FIOH), p. 185.

Guralnik, Jack M., Andre Z. v, Robert D. Abbott, Lisa F. Berkman, Suzanne Satterfield, Denis A. Evans, and Robert B. Wallace. 1993. Maintaining mobility in life. I. Demographic characteristics and chronic conditions. *American Journal of Epidemiology* 137: 845–57. [CrossRef]

Hinrichs, Timo, Mikaela B. von Bonsdorff, Timo Törmäkangas, Monika E. von Bonsdorff, Jenni Kulmala, Jorma Seitsamo, Clas-Håkan Nygård, Juhani Ilmarinen, and Taina Rantanen. 2014. Inverse effects of midlife occupational and leisure time physical activity on mobility in old age—A 28-year prospective follow-up study. *Journal of the American Geriatrics Society* 62: 812–20. [CrossRef]

Ilmarinen, Juhani. 2009. Work Ability—A comprehensive concept for occupational health research and prevention. Editorial. *Scandinavian Journal of Work, Environment & Health* 35: 1–5.

Ilmarinen, Juhani. 2013. Redesign of workplaces for an ageing society. In *Older Workers in an Ageing Society: Critical Topics in Research and Policy*. Edited by Philip Taylor. Cheltenham: Edward Elgar Publishing, pp. 133–46.

Ilmarinen, Juhani. 2019. From work ability research to implementation. *International Journal of Environmental Research and Public Health* 16: 2882. [CrossRef] [PubMed]

Ilmarinen, Juhani, and Jorma Rantanen. 1999. Promotion of work ability during ageing. *American Journal of Industrial Medicine* 36: 21–23. [CrossRef]

Ilmarinen, Juhani, Timo Suurnäkki, Clas-Håkan Nygård, and Kurt Landau. 1991. Classification of municipal occupations. *Scandinavian Journal of Work, Environment & Health* 17: 12–29.

Ilmarinen Juhani, Tuomi Kaija, and Klockars Matti. 1997. Changes in work ability of active employees over an 11-year period. *Scandinavian Journal of Work, Environment & Health* 23: 49–57.

Ilmarinen, Juhani, Kaija Tuomi, and Jorma Seitsamo. 2005. *New Dimensions of Work Ability*. International Congress Series; Amsterdam: Elsevier, vol. 1280, pp. 3–7.

Kulmala, Jenni, Timo Hinrichs, Mikaela B. von Bonsdorff, Monika E. von Bonsdorff, Clas-Håkan Nygård, Matti Klockars, Jorma Seitsamo, Juhani Ilmarinen, and Taina Rantanen. 2014. Work-related Stress in Midlife Is Associated with Higher Number of Mobility Limitation in Old Age—Results from the FLAME Study. *Age* 36: 9722. [CrossRef]

Mein, Gill, Pekka Martikainen, Harry Hemingway, Stephen Stansfeld, and Michael Marmot. 2003. Is retirement good or bad for mental and physical health functioning? Whitehall II longitudinal study of civil servants. *Journal of Epidemiology and Community Health* 57: 46–49. [CrossRef]

Neupane, Subas, Clas-Håkan Nygård, K. C. Prakash, Mikaela B. von Bonsdorff, Monika E. von Bonsdorff, Jorma Seitsamo, Taina Rantanen, Juhani Ilmarinen, and Päivi Leino-Arjas. 2018. Multisite musculoskeletal pain trajectories from midlife to old age: A 28-year follow-up of municipal employees. *Occupational and Environmental Medicine* 75: 863–70. [CrossRef]

Nylund, Karen L., Tihomir Asparouhov, and Bengt O. Muthén. 2007. Deciding on the number of classes in Latent Class Analysis and Growth Mixture Modelling: A Monte Carlo simulation study. *Structural Equation Modeling* 14: 535–69. [CrossRef]

Oakman, Jodi, Subas Neupane, and Clas-Håkan Nygård. 2016. Does age matter in predicting musculoskeletal disorder risk? An analysis of workplace predictors over 4 years. *International Archives of Occupational and Environmental Health* 89: 1127–36. [CrossRef] [PubMed]

Oakman, Jodi, Subas Neupane, K. C. Prakash, and Clas-Håkan Nygård. 2019. What Are the Key Workplace Influences on Pathways of Work Ability? A Six-Year Follow Up. *International Journal of Environmental Research and Public Health* 16: 2363. [CrossRef] [PubMed]

Prakash, K.C., Subas Neupane, Päivi Leino-Arjas, Mikaela B. von Bonsdorff, Taina Rantanen, Monika E. von Bonsdorff, Jorma Seitsamo, Juhani Ilmarinen, and Clas-Håkan Nygård. 2016. Midlife job profiles and disabilities in later life: A 28-year follow-up of municipal employees in Finland. *International Archives of Occupational and Environmental Health* 89: 997–1007. [CrossRef] [PubMed]

Prakash, K.C., Subas Neupane, Päivi Leino-Arjas, Mikaela B. von Bonsdorff, Taina Rantanen, Monika E. von Bonsdorff, Jorma Seitsamo, Juhani Ilmarinen, and Clas-Håkan Nygård. 2017. Work-related biomechanical exposure and job strain as separate and joint predictors of musculoskeletal diseases: A 28-year prospective follow-up study. *American Journal of Epidemiology* 186: 1256–67. [CrossRef] [PubMed]

Prakash, K.C., Subas Neupane, Päivi Leino-Arjas, Mikko Härmä, Mikaela B. von Bonsdorff, Taina Rantanen, Monika E. von Bonsdorff, Jorma Seitsamo, Timo Henrichs, Juhani Ilmarinen, and et al. 2019. Trajectories of mobility limitations over 24 years and their characterization by shift work and leisure-time physical activity in midlife. *European Journal of Public Health* 29: 882–88. [CrossRef] [PubMed]

Ram, Nilam, and Kevin J. Grimm. 2009. Growth mixture modeling: A method for identifying difference in longitudinal change among unobserved groups. *International Journal of Behavioral Development* 33: 565–76. [CrossRef]

Robertson, Raymond, Hongyang Di, Drusilla Brown, and Rajeev H. Dehejia. 2016. *Working Conditions, Work Outcomes, and Policy in Asian Developing Countries*. NYU Wagner Research Paper, (2856292). Manila: Asian Development Bank.

Sluiter, Judith K. 2006. High-demand jobs: Age-related diversity in work ability? *Applied Ergonomics* 37: 429–40. [CrossRef] [PubMed]

Stenholm, Sari, Mika Kivimäki, Marja Jylhä, Ichiro Kawachi, Hugo Westerlund, Jaana Pentti, Marcel Goldberg, Marie Zins, and Jussi Vahtera. 2016. Trajectories of self-rated health in the last 15 years of life by cause of death. *European Journal of Epidemiology* 31: 177–85. [CrossRef] [PubMed]

Sörensen, Lars E., Mika M. Pekkonen, Kaisa H. Männikkö, Veikko A. Louhevaara, Juhani Smolander, and Markku J. Alén. 2008. Associations between work ability, health-related quality of life, physical activity and fitness among middle-aged men. *Applied Ergonomics* 39: 786–91. [CrossRef] [PubMed]

Tuomi, Kaija, Juhani Ilmarinen, Matti Klockars, Clas-Håkan Nygård, Jorma Seitsamo, Pekka Huuhtanen, Rami Martikainen, and Lea Aalto. 1997. Finnish research project on aging workers in 1981–1992. *Scandinavian Journal of Work, Environment & Health* 23: 7–11.

Tuomi, Kaija, Ilmarinen Juhani, Antti Jahkola, Lea Katajarinne, and Arto Tulkki. 1998. *Work Ability Index*, 2nd ed. Helsinki: Finnish Institute of Occupational Health.

United Nations. 2015. Sustainable Development Goals. Available online: http://www.un.org/sustainabledevelopment/sustainable-development-goals/ (accessed on 5 January 2020).

Van Dam, Karen, Tinka van Vuuren, and Sofie Kemps. 2017. Sustainable employment: The importance of intrinsically valuable work and an age-supportive climate. *The International Journal of Human Resource Management* 28: 2449–72. [CrossRef]

Van Vuuren, Tinka, and Karen Van Dam. 2013. Sustainable employability by vitalizing: The importance of development opportunities and challenging work for workers' sustainable employability. In *Life-Long Employability? Perspectives on Sustainable Employment at Work: Interventions and Best Practices*. Edited by Annet De Lange and Beatrice I. J. M. Van der Heijden. Vakmedianet: Alphen aan de Rijn, pp. 357–76.

Verbrugge, Lois M., and Alan M. Jette. 1994. The disablement process. *Social Science & Medicine* 38: 1–14.

von Bonsdorff, Mikaela B., Jorma Seitsamo, Juhani Ilmarinen, Clas-Håkan Nygård, Monika E. von Bonsdorff, and Taina Rantanen. 2011. Work ability in midlife as a predictor of mortality and disability in later life: A 28-year prospective follow-up study. *CMAJ* 183: e235–42. [CrossRef] [PubMed] von Bonsdorff, E. Monika, Taina Rantanen, Timo Törmäkangas, Jenni Kulmala, Timo Hinrichs, Jorma Seitsamo, Clas-Håkan Nygård, Juhani Ilmarinen, and Mikaela B. von Bonsdorff. 2016. Midlife work ability and mobility limitation in old age among non-disability and disability retirees—A prospective study. *BMC Public Health* 16: 154. [CrossRef] [PubMed]

Westerlund, Hugo, Mika Kivimäki, Anita Singh-Manous, Maria Melchior, Jane E. Ferrie, Jaana Pentti, Markus Jokela, Constanze Leineweber, Marcel Goldberg, Marie Zins, and Jussi Vahtera. 2009. Self-rated health before and after retirement in France (GAZEL): A cohort study. *The Lancet* 374: 1889–96. [CrossRef]

World Health Organization. 2001. *International Classification of Functioning, Disability and Health (ICF)*. Geneva: WHO.

MDPI

St. Alban-Anlage 66

4052 Basel

Switzerland

Tel. +41 61 683 77 34

Fax +41 61 302 89 18

www.mdpi.com

MDPI Books Editorial Office

E-mail: books@mdpi.com

www.mdpi.com/books